"十二五"职业教育国家规划教材

经全国职业教育教材审定委员会审定

（修订版）

路由型与交换型互联网基础实训手册

第4版

主　编　杨鹤男　张　鹏
副主编　闫立国　何　琳
参　编　包　楠　李晓隆　罗　忠　沈天珞
　　　　闫昊伸　吴翰青　石　柳　周志荣
　　　　包清太　赵传兴

机械工业出版社

本书是"十二五"职业教育国家规划教材，是在第3版的基础上修订而成的。

本书是神州数码DCNE认证的指定教材，全书共4个部分，分别为交换机实训、路由器实训、无线实训和防火墙实训，内容涉及网络工程师在实际工作中遇到的各种典型问题的主流解决方案及实施步骤。

本书可作为高等职业院校计算机应用专业和网络技术应用专业的实训教材，也可作为从事交换机、路由器管理和网络维护有关人员的工作指导书，还可作为计算机网络工程技术岗位培训的实训教材。

本书配有微课视频，读者用手机扫一扫书中二维码，即可观看。

为便于教学，本书配有电子课件，选择本书作为教材的教师可来电(010-88379194)索取，或登录网站www.cmpedu.com，注册后免费下载。

图书在版编目（CIP）数据

路由型与交换型互联网基础实训手册/杨鹤男，张鹏主编．—4版．—北京：机械工业出版社，2021.4（2025.1重印）

"十二五"职业教育国家规划教材：修订版

ISBN 978-7-111-67811-3

Ⅰ．①路… Ⅱ．①杨… ②张… Ⅲ．①互联网络—路由选择—高等职业教育—教材 Ⅳ．①TN915.05

中国版本图书馆CIP数据核字（2021）第051354号

机械工业出版社（北京市百万庄大街22号 邮政编码100037）

策划编辑：李绍坤 梁 伟　　责任编辑：李绍坤 刘益汛

责任校对：马立婷　　封面设计：鞠 杨

责任印制：常天培

北京机工印刷厂有限公司印刷

2025年1月第4版第8次印刷

184mm×260mm・12.75 印张・302千字

标准书号：ISBN 978-7-111-67811-3

定价：42.00元

电话服务　　　　　　　　　网络服务

客服电话：010-88361066　　机　工　官　网：www.cmpbook.com

　　　　　010-88379833　　机　工　官　博：weibo.com/cmp1952

　　　　　010-68326294　　金　书　网：www.golden-book.com

封底无防伪标均为盗版　　机工教育服务网：www.cmpedu.com

前言

本书是"十二五"职业教育国家规划教材，是在第3版的基础上修订而成的。

本书是神州数码DCNE（神州数码认证网络工程师）认证考试的指定学习教材，书中首先简明地介绍了交换型网络的实训案例，然后对路由型网络的实训案例进行了详细的阐述，最后讨论了无线网络的实训案例和网络安全的实训案例。

全书共4个部分，分别为交换机实训、路由器实训、无线实训和防火墙实训。内容涉及网络工程师在实际工作中遇到的各种典型问题的主流解决方案及实施步骤。

本书在修订过程中努力体现以下特色：

1. 融入思政元素，立德树人

本书所讲述的技术和引用的案例都是神州数码推荐的设计方案和典型的成功案例，学生在学习过程中不仅能学到知识和技术，还能认识到在实际工作中必须具有的严谨、认真的工作态度，培养学生的社会责任感。

2. 校企"双元"团队编写，产教融合

本书由校企"双元"团队编写，编写团队包括北京神州数码云科信息技术有限公司的杨鹤男、闫立国、包楠、李晓隆、张鹏，北京市求实职业学校的何琳、沈天珞、闫昊伸、吴瀚青，深圳市第二职业技术学院的罗忠，湖北生物科技职业学院的石柳，云南临沧职业技术学院的周志荣，广州市增城区广播电视大学的包清太，以及甘肃财贸职业学院的赵传兴。

3. 创新编写体例，任务驱动

本书采用任务驱动模式进行编写，每章的实训任务都来自企业真实案例，经过精心提炼，具有很强的针对性、实用性。每个实训任务以"实训目标——实训拓扑——实训任务"的结构进行组织，这样，读者不仅能通过任务实施完成书中所介绍的任务，还能建立通用拓扑体系，举一反三，应用所学于实际工作项目中。

4. 配套资源丰富

本书配有对应实训操作的微课视频，读者可以使用手机扫描书中二维码直接观看。

同时，配套面向教师的电子课件等教学资源，支持传统教学与混合式教学、移动式教学等信息化教学模式的融合创新。

本书由杨鹤男、张鹏任主编，闫立国、何琳任副主编。参与编写的还有包楠、李晓隆、罗忠、沈天琀、闫昊伸、吴翰青、石柳、周志荣、包清太和赵传兴。本书的第1、2章由闫立国、包楠编写，第3、4章由张鹏、罗忠、周志荣编写，第5、6章由石柳、包清太、赵传兴编写，第7～9章由杨鹤男、何琳、沈天琀编写，第10、11章由闫昊伸、吴翰青编写，李晓隆负责内容核对。

● 关于图标：

本书图标采用了神州数码图标库标准图标，除真实设备外，所有逻辑示意均使用如下图标。

由于编者水平有限，书中不妥之处在所难免，恳请读者批评指正。

编　者

二维码索引

序号	视频名称	二维码	页码	序号	视频名称	二维码	页码
1	交换机管理		2	8	OSPF 路由协议		52 113
2	VLAN 与 TRUNK		17 20	9	ACL 访问控制列表		56 61
3	RSTP 与 MSTP		23 29	10	路由器设备简介		76
4	链路聚合		34	11	NAT 网络地址转换		133
5	二层安全技术		37	12	VPN 工作原理及配置		136 143
6	路由技术与应用		44 97	13	无线技术基础		166
7	RIP 路由协议		48 107	14	防火墙		176

目 录

前言
二维码索引

第 1 部分 交换机实训

第 1 章 交换机实践基础 2
- 实训 1 交换机认识与带外管理 2
- 实训 2 交换机配置模式与 CLI 调试 5
- 实训 3 管理交换机配置文件 10
- 实训 4 交换机带内管理 13

第 2 章 交换机典型园区应用实践 17
- 实训 1 单台交换机 VLAN 17
- 实训 2 跨交换机 VLAN 20
- 实训 3 生成树协议 23
- 实训 4 多实例生成树协议 29
- 实训 5 链路聚合 34
- 实训 6 端口安全 37

第 3 章 交换机路由应用实践 44
- 实训 1 单臂路由实现 VLAN 间互访 44
- 实训 2 三层交换机实现 VLAN 间互访 46
- 实训 3 动态路由协议 RIP 48
- 实训 4 动态路由协议 OSPF 52

第 4 章 交换机高级应用实践 56
- 实训 1 标准访问控制列表 56
- 实训 2 扩展访问控制列表 61
- 实训 3 三层交换机 DHCP 服务 65
- 实训 4 三层交换机 DHCP 中继 67
- 实训 5 虚拟路由器冗余协议 70

第 2 部分 路由器实训

第 5 章 路由器实践基础 76
- 实训 1 路由器的基本管理方法 76
- 实训 2 维护路由器的配置文件 82

实训 3　路由器直连路由的配置 *90*
　　实训 4　路由器单臂路由的配置 *93*
第 6 章　路由器路由技术基础 *97*
　　实训 1　路由器静态路由的配置 *97*
　　实训 2　静态路由掩码最长匹配 *102*
　　实训 3　路由器 RIP 的配置 *107*
　　实训 4　路由器单区域 OSPF 协议的配置 *113*
第 7 章　路由器接口设置实践 *118*
　　实训 1　路由器串口 PPP-PAP 配置 *118*
　　实训 2　路由器串口 PPP-CHAP 配置 *122*
第 8 章　路由器应用技术实践 *126*
　　实训 1　标准访问列表配置 *126*
　　实训 2　扩展访问列表配置 *131*
　　实训 3　路由器 NAT 实训 *133*
　　实训 4　IPSec VPN（IKE）的配置 *136*
　　实训 5　L2TP/PPTP VPN 的配置 *143*
第 9 章　路由器综合应用进阶 *148*
　　实训 1　综合实验 1 .... *148*
　　实训 2　综合实验 2 .... *155*

第 3 部分　无线实训

第 10 章　无线技术基础 *166*
　　实训 1　多 SSID 的无线接入配置 *166*
　　实训 2　开放式无线接入配置 *169*
　　实训 3　无线加密接入的配置 *171*

第 4 部分　防火墙实训

第 11 章　防火墙技术基础 *176*
　　实训 1　防火墙透明模式的初始配置 *176*
　　实训 2　防火墙路由模式的初始配置 *179*
　　实训 3　防火墙混合模式的初始配置 *183*
　　实训 4　防火墙 SNAT 配置 *187*
　　实训 5　防火墙 DNAT 配置 *190*

参考文献 *195*

第 1 部分

交换机实训

第1章 交换机实践基础

实训1 交换机认识与带外管理

实训目标

交换机管理

小张中职毕业后在朋友开的网吧里当网管,后经熟人推荐来到某公司做网络中心的网络管理员。由于之前没有使用过网管型交换机,所以他的师傅李工程师(以下简称其"李工")给他交代了以下3个任务目标。同时李工告诉小张网络强国是国家战略,包括网络基础设施建设、信息通信业新的发展和网络信息安全三个方面。网络是基础,只有打好基础才能为国家战略发挥余热!

1)熟悉网管型二层交换机的外观。
2)了解交换机各端口的名称和作用。
3)了解交换机最基本的管理方式——带外管理的方法。

通过学习本实训,熟悉交换机的外观,了解交换机各端口名称和作用,具备带外管理的能力。

实训拓扑

实训拓扑图如图1-1所示。

图 1-1

实训任务

步骤1:观察交换机外观,认识交换机Console口及各网口的编号规则,如图1-2所示。

图 1-2

小张仔细观察了这种网管型交换机，通过向李工咨询，了解了交换机各端口的作用，知道了 Console 口就是用于带外管理的配置口，可连接在 PC 的串行口上对交换机进行管理。同时，知道了以太网口通常用三位数字表示：0/0/1 中的第一个 0 表示堆叠中的第一台交换机，如果是 1，就表示第 2 台交换机；第 2 个 0 表示交换机上的第 1 个模块，而第 3 位数字则表示端口号。

例如，0/0/1 表示的是堆叠中第 1 台交换机第 1 个网络端口模块上的第 1 个网络端口。

默认情况下，如果不存在堆叠，则交换机总会认为自己是第 0 台交换机。

提示：交换机的端口命名有 2 段式（x/x）和 3 段式（x/x/x）两种，具体的端口命名格式可以通过登录设备后使用 show running-config 命令查看。

步骤 2：在关闭电源的情况下用 Console 线连接交换机与 PC。

小张观察了 Console 线，发现一端是 RJ-45 水晶头，另一端则是 RS-232 串行接口，于是他将 RJ-45 水晶头插入交换机的 Console 口，另一端插入 PC 的 RS-232 串口。

步骤 3：使用 CRT 工具登录交换机 CLI 管理界面。

小张打开了交换机与 PC 的电源，通过咨询李工，了解了接入方法。PC 启动完成后，找到 SecureCRT 工具，打开 SecureCRT。

单击"File"→"Quick Connect"打开快速连接窗口，如图 1-3 所示。

图 1-3

李工告诉小张 CRT 是网络工程必备工具，他可以配置 SSH、Telnet、Serial（串口）等连接到交换机、路由器、服务器等设备。

这里通过 Serial（串口）连接到交换机，选择"Serial"，如图 1-4 所示。

图 1-4

李工告诉小张可以在"设备管理器"的"端口（COM 和 LPT）"中查看端口，本例是 COM3，如图 1-5 所示。

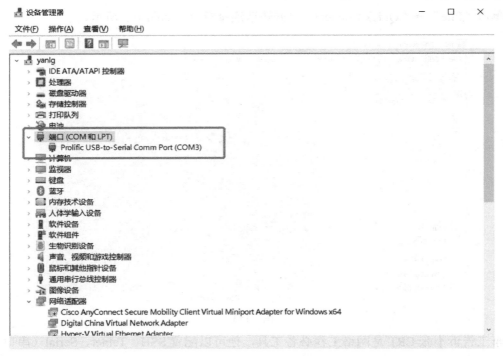

图 1-5

因此，在图 1-6 所示界面中，Port 就选择"COM3"。

第 1 章　交换机实践基础

图　1-6

交换机的波特率是"9600"，数据位选"8"，奇偶校验选"None"，停止位选"1"，数据流控制选"无"，单击"连接"按钮就可以登录交换机了。

实训 2　交换机配置模式与 CLI 调试

实训目标

李工告诉小张，CLI（Command Line Interface，命令行界面）和 GUI（Graphical User Interface，图形界面）相对应。CLI 由 Shell 程序提供，是由一系列的配置命令组成的。根据这些命令在配置管理交换机时所起的作用不同，Shell 将这些命令分类，不同类别的命令对应着不同的配置模式。

CLI 是交换机调试界面中的主流界面，基本上所有的网络设备都支持命令行界面。国内外主流的网络设备供应商使用很相近的命令行界面，方便用户调试不同厂商的设备。神州数码网络产品的调试界面兼容国内外主流厂商的界面，和思科命令行接近，便于学习。只有少部分厂商使用自己独有的配置命令。

通过学习本实训，熟悉交换机各配置模式，具备 CLI 调试的能力。

实训拓扑

实训拓扑图如图 1-7 所示。

5

图 1-7

实训任务

步骤1：一般用户配置模式

观察实训1出现的CLI，发现提示符由两部分组成，前面是switch，这个名称可以改变；后面跟着一个大于号">"，这个">"号表示当前交换机处于一般用户配置模式。之所以要把它称为一般用户配置模式，是因为任何人通过Console口接入交换机都可以进入这个模式。

在一般用户配置模式下，可用的命令比较少，输入"？"后即可看到该模式下可用的所有命令，如图1-8所示。

图 1-8

步骤2：特权用户配置模式。

在一般用户配置模式下，输入命令"enable"并按<Enter>键，可以看到提示符变成了"#"，如图1-9所示。

提示符"#"表示当前交换机处于特权用户配置模式。在特权用户配置模式下，可以查询交换机配置信息、各个端口的连接情况、收发数据统计等。而且进入特权用户配置模式后，可以进入全局模式对交换机的各项配置进行修改，因此进行特权用户配置模式必须设置特权用户密码，防止非特权用户非法使用，对交换机配置进行恶意修改，造成不必要的损失。

要从特权配置模式返回一般用户配置模式，可以使用"exit"命令。

图 1-9

步骤3：全局配置模式。

在特权用户配置模式下，使用"config terminal"命令可以进入全局配置模式。在全局配置模式中，可以对交换机进行全局性的配置，如配置 MAC 地址表、端口镜像、创建 VLAN、启动 IGMP Snooping、STP 等。在全局模式下还可以通过命令进入端口对各个端口进行配置。

在全局配置模式中配置特权用户的密码，代码如下。

switch>enable

switch#config terminal

switch(config)#enable password digitalchina

switch(config)#

返回一般用户配置模式进行验证，代码如下。

switch(config)#exit

switch#exit

switch>enable

Password:

switch#config terminal

switch(config)#

第一个 exit 是从全局配置模式返回特权用户配置模式，第二个 exit 是从特权用户配置模式返回一般用户配置模式。然后输入命令 enable，出现密码提示，输入刚才创建的特权密码 digitalchina 即可进入特权用户配置模式。再输入 config terminal 便进入了全局配置模式，如图 1-10 所示。

图 1-10

步骤4：熟悉"？"的使用。

前面熟悉了三种配置模式，在全局配置模式中还有端口模式、VLAN 模式等子模式，这些在后面的任务中再介绍。现在，先来熟悉一下"？"符号。

在实训1中已经接触过了"？"，在">"提示符下输入"？"便列出了该模式下的所有命令。

这个方法可以用在任何模式中。例如，在特权用户配置模式下输入"？"，屏幕显示如图 1-11 所示。

图 1-11

从图 1-11 看到列出了很多特权用户配置模式下的命令。可以按空格键继续显示下一屏，也可以按 <Enter> 键继续显示下一行。

除此之外，还有两种场合可以使用"？"。

第一种，如果只记住了某命令的前几个字母，则可以使用"？"查询，代码如下。

switch#lan?
   language   设置语言
switch#co?
   config   进入配置模式
   copy   复制文件

查询结果表示在特权模式下以 lan 开头的命令只有一个，而以 co 开头的命令有两个。

第二种，绝大部分命令后需要参数，可以用"？"来查询参数，代码如下。

switch(config)#hostname ?
   WORD   主机名 <0-30> 字符 (符号只允许使用字母、数字和下划线)
   <cr>

在命令 hostname 后按空格键后跟一个"？"，可以看到该命令需跟一个主机名，也可以直接按 <Enter> 键，代码如下。

switch(config)#hostname switch ?
   <cr>
switch(config)#hostname switch
switch(config)#

输入 hostname switch 后按空格键和"？"，看到没有参数了，可以直接按 <Enter> 键然后提示符主机名部分被更改。

步骤 5：<Tab> 键的使用。

学会使用"？"，就可以不用去记忆繁多的命令了。但 CLI 还提供了更方便快捷的 <Tab> 键。如果一个命令的前几个字符没有歧义，则可以使用 <Tab> 键来补全命令的输入，代码如下。

```
switch#show ve
% Ambiguous command: "show ve"        // 只输入 ve 不行，有歧义
switch#show ver                        // 输入 ver 后按 <Tab> 键，可以补全命令
  CS6200 Device, Compiled on Jul 26 16:27:26 2010
  SoftWare Version 6.1.73.13
  BootRom Version 3.0.12
  HardWare Version R01
  Device serial number A830004862
  Copyright (C) 2001-2009 by Digital China Networks Limited.
  All rights reserved
  Uptime is 0 weeks, 0 days, 7 hours, 32 minutes
```

只有当前命令正确的情况下才可以使用 <Tab> 键，也就是说，一旦命令没有输入完全，但是 <Tab> 键又没有起作用时，就说明当前的命令中出现了错误，或者命令错误，或者参数错误等，需要仔细排查。

步骤 6：命令的不完全匹配。

<Tab> 键可以帮助我们熟悉命令，而熟练之后，则一般很少使用 <Tab> 键了。

绝大多数情况下，能使用 <Tab> 键的地方都可以省略它。

例如，显示版本号的命令，完全命令是"show version"，使用 <Tab> 键的代码如下。

```
switch#sh<Tab> ver<Tab>
```

如果把 <Tab> 键省略，则变成如下代码。

```
switch#sh ver
```

这就是命令的不完全匹配。

步骤 7：上下光标键"↑""↓"的使用。

当输入、执行了一些命令后，可以使用上下光标键"↑""↓"来浏览已执行过的命令，当需要重复执行相似命令时，可以大大节省时间。

步骤 8：查看错误信息。

第一种错误信息是"%Ambiguous command: "show ve""，表示命令"show ve"有歧义——可能有两种以上的解释。遇到这种情况，再多输入一个或两个字母，就可以避免这种情况。

第二种错误信息是把命令写错了，代码如下。

```
switch>sh valn
             ^
% Invalid input detected at '^' marker.
```

上例把"vlan"写成了"valn"，出现了错误信息，并用"^"标示了出错的地方。

第三种错误信息是没有跟命令参数，代码如下。

```
switch#show
% Incomplete command.
```

上例表示命令不完整。

实训3　管理交换机配置文件

实训目标

小张：师傅，如果交换机配置好了，重启交换机会怎么样呢？配置好的信息保存在哪里？

李工：和我们使用计算机一样，如果没有保存过，重启后配置信息也会丢失。当交换机应用环境发生改变时，就需要清空交换机配置，再重新对交换机进行配置以适应新的应用环境，还需要将已配置的信息进行保存。

通过学习本实训，了解交换机配置文件的使用，具备管理交换机配置文件的能力。

实训拓扑

实训拓扑图如图1-12所示。

图　1-12

实训任务

步骤1：清空交换机配置，代码如下。

```
switch>enable                                    //进入特权用户配置模式
Password:                                        //输入特权密码 digitalchina
switch#set default                               //使用 set default 命令
Are you sure? [Y/N] = y                          //是否确认？
switch#write                                     //清空 startup-config 文件
switch configuration has been set default!
switch #show startup-config                      //显示当前的 startup-config 文件
% Current startup-configuration is default factory configuration!   //系统提示此启动文件为出厂默认配置
switch#reload                                    //重新启动交换机
Process with reboot? [Y/N]y
```

小张：这样做了之后，交换机的所有配置信息都清除了吗？特权密码也清除了吗？

李工：是的，你可以试试看。

```
switch>enable
switch#
```

小张：果然不需要密码了。

李工：交换机已恢复到出厂默认配置，所有后来做的配置都清除了。

步骤2：配置交换机日期时间，代码如下。

```
switch#clock set ?                          // 使用？查询命令格式
  HH:MM:SS  Hour:Minute:Second
switch#clock set 9:52:30                    // 配置当前时间
Current time is Sun Jan 01 09:52:30 2006    // 配置完即有显示，但年份不对
switch#clock set 9:52:30 ?                  // 使用？查询，原来命令没有结束
  YYYY.MM.DD  Year.Month.Day(Valid time is between 1970.1.1 and 2038.12.31)
  <cr>
switch#clock set 9:52:30 2012.7.31          // 配置当前日期和时间
Current time is Tue Jul 31 09:52:30 2012    // 显示正确，配置完成
switch#
```

步骤 3：查看交换机闪存中的文件，代码如下。

```
switch#show flash

nos.img                          4,702,429
startup-config                   24

Used    4,702,453 bytes in 2 files, Free   3,686,155 bytes.
switch#
```

可以看到在交换机闪存中保存了两个文件，一个是交换机操作系统文件 nos.img，另一个是配置文件 startup-config。下面来备份这两个文件。

步骤 4：配置 TFTP 服务器。

首先，需要把 PC 配置成 TFTP 服务器。TFTP（Trivial File Transfer Protocol，微型文件传输协议）与 FTP 不同，TFTP 采用 UDP 进行文件传输，UDP 在传输时使用的是不可靠的数据流传输服务，同时也不支持用户认证机制以及设置用户对文件操作的授权；它是通过发送包文，应答方式，加上超时重传方式来保证数据的正确传输。TFTP 相对于 FTP 的优点是提供简单的、开销不大的文件传输服务。

市场上 TFTP 服务器的软件很多，各种软件虽然界面不同，但功能都是一样的，使用方法也都类似：首先是 TFTP 软件安装，安装完毕之后设定根目录，需要使用的时候，开启 TFTP 服务器即可。下面以 Tftpd32 为例来配置 TFTP 服务器。

双击 Tftpd32.exe 文件，出现 TFTP 主界面，如图 1-13 所示。

图　1-13

在主界面中可以看到该服务器的根目录是 D:\BackupSwitch,服务器的 IP 地址 192.168.1.10 也自动出现在 Server interfaces 下拉列表中。可以通过单击"设置"按钮来更改 TFTP 服务器的根目录。

步骤 5:配置交换机管理 IP 地址。

交换机管理 VLAN 默认为 VLAN 1,可以通过配置 VLAN 1 的 IP 地址实现 PC 网口与交换机以太网口的通信。代码如下。

```
switch#config terminal                              // 进入全局配置模式
switch(config)#interface vlan 1                     // 进入 VLAN 1 配置接口
switch(config-if-vlan1)#ip address 192.168.1.1 255.255.255.0
                                                    // 配置 VLAN 1 的 IP 地址与掩码
switch(config-if-vlan1)#no shutdown                 // 打开 VLAN 1 接口
switch(config-if-vlan1)#exit                        // 返回全局配置模式
switch(config)#
```

步骤 6:验证 TFTP 服务器与交换机的连通性,代码如下。

```
switch#ping 192.168.1.10                            // 特权用户配置模式下 ping TFTP 服务器 IP
Type ^c to abort.
Sending 5 56-byte ICMP Echos to 192.168.1.10, timeout is 2 seconds.
!!!!!                                               // 出现 5 个 "!" 号,表示 ping 通了
Success rate is 100 percent (5/5), round-trip min/avg/max = 0/3/16 ms
switch#
```

步骤 7:备份交换机配置文件。

将步骤 3 中看到的交换机配置文件备份到 TFTP 服务器中,代码如下。

```
switch#write                                        // 保存当前配置到 startup-config 中
Write configuration successfully!
switch#copy startup-config tftp://192.168.1.10/sc2012.7
                    // 将配置文件 startup-config 上传到 TFTP 服务器根目录中,并改名为 sc2012.7
Confirm copy file [Y/N]:y
Begin to send file, please wait...


File transfer complete.
close tftp client.
switch#copy nos.img tftp://192.168.1.10
                    // 将交换机操作系统文件上传到 TFTP 服务器根目录中,不改名
Confirm copy file [Y/N]:y
Begin to send file, please wait...
#########################################################################
#################################################
File transfer complete.
close tftp client.
switch#
```

步骤 8：下载配置文件，代码如下。

```
switch#copy tftp://192.168.1.10/sc2012.7 startup-config
                                         // 将备份文件 sc2012.7 恢复成交换机启动配置文件 startup-config
Confirm to overwrite the existed destination file?    [Y/N]:y
Begin to receive file, please wait...

File transfer complete.
Recv total 907 bytes
Write ok.
close tftp client.
switch#reload                            // 重启交换机
```

步骤 9：升级交换机操作系统。

下载交换机操作系统升级版本到 TFTP 服务器根目录，代码如下。

```
switch#copy tftp://192.168.1.10/nos.img nos.img
Confirm to overwrite the existed destination file?  [Y/N]:y
Begin to receive file, please wait...
Get Img file size success, Img file size is:4702429(bytes).
################################################################################################
###############################
File transfer complete.
Recv total 4702429 bytes

Write ok.
close tftp client.
switch#
```

实训 4　交换机带内管理

实训目标

小张：师傅，如果一个局域网内有很多交换机分别放置在不同的位置，这样一台一台地去配置不是很麻烦吗？

李工：我们前面学习的是交换机的带外管理方式，也就是使用交换机的 Console 口去管理。而网络是互通的，在网络的任何一个信息点都能访问其他的信息点，为什么不通过网络的方式来管理交换机呢？通过带内管理方式，管理员就可以在办公室中调试网络中所有的交换机了。

带内管理方式有两种，分别是 Telnet 方式和 Web 方式。

但是，如果交换机的某些配置或错误导致带内管理方式失效，就必须使用带外管理对交换机进行配置管理。

通过学习本实训，了解交换机带内管理方式，具备配置交换机带内管理的能力。

实训拓扑

实训拓扑图如图 1-14 所示。

图 1-14

实训任务

步骤 1：通过 Console 口配置交换机的基本信息。

在使用带内管理方式前，必须对交换机做一些基本配置，代码如下。

```
// 配置特权用户模式密码为 digitalchina
switch>enable
switch#config terminal
switch(config)#enable password digitalchina
switch(config)#


// 配置用户名认证，用户名为 dcs，权限为 15，密码为 digital
switch(config)#username dcs privilege 15 password digital
switch(config)#


// 开启 Telnet 服务和 Web 服务
switch(config)#telnet-server enable
Telnet server has been already enabled.          //Telnet 服务默认是开启的
switch(config)#ip http server
web server is on
switch(config)#


// 配置管理 VLAN 的 IP 地址
switch(config)#interface vlan 1
switch(config-if-vlan1)#ip address 192.168.1.1 255.255.255.0
switch(config-if-vlan1)#exit
switch(config)#
```

步骤 2：验证 PC 与交换机的连通性，代码如下。

```
switch#ping 192.168.1.10
Type ^c to abort.
Sending 5 56-byte ICMP Echos to 192.168.1.10, timeout is 2 seconds.
!!!!!                       // 出现 5 个 "!"，表示交换机与 PC 通信正常
```

Success rate is 100 percent (5/5), round-trip min/avg/max = 0/0/0 ms
switch#

步骤 3：使用 Telnet 方式管理交换机。

此时，可以拔除 Console 连接。执行"开始"→"运行"命令，在"运行"文本框中输入"cmd"，启动命令提示符窗口，输入"telnet 192.168.1.1"，使用 Telnet 方式登录交换机，如图 1-15 所示。

图 1-15

输入用户名 dcs，密码 digital，便进入了交换机特权用户配置模式（用户 dcs 权限等级为最高级 15，直接进入特权用户配置模式），如图 1-16 所示。

图 1-16

步骤 4：使用 Web 方式管理交换机。

打开浏览器，在地址栏输入"http://192.168.1.1"，便出现了交换机的 Web 登录界面，如图 1-17 所示。

图 1-17

输入用户名 dcs，密码 digital，进入管理界面，如图 1-18 所示。

图 1-18

Web 管理方式比较简单，如果不习惯于 CLI 调试，则可以采用 Web 方式调试。但主流的调试界面还是 CLI，所以还是要以学习 CLI 为主。

第 2 章 交换机典型园区应用实践

实训1 单台交换机 VLAN

VLAN 与 TRUNK

实训目标

李工：现在我们开始要进入园区网的应用实践，首先要接触 VLAN。小张，还记得什么是 VLAN 以及为什么要划分 VLAN 吗？

小张：VLAN 是虚拟局域网的意思，划分 VLAN 的主要目的是隔离广播域。

李工：对。由于交换机所有端口属于同一个广播域，而网络中存在大量的广播包，VLAN 技术能够有效地进行隔离，从而使网络性能得到提升。

例如，某学校的两个机房位于同一楼层，所有信息端口都连接在一台交换机上。为了保证两个机房的相对独立，不影响各机房内部的通信效率，保证两个机房之间互不干扰，就需要划分 VLAN，使这两个机房处于不同的 VLAN 内。

通过学习本实训，能够掌握 VLAN 的概念，具备配置交换机 VLAN 的能力。

实训拓扑

实训拓扑图如图 2-1 所示。

图 2-1

实训任务

步骤 1：规划交换机的 VLAN 及端口，分配 PC 的 IP 地址，见表 2-1。

表 2-1

VLAN	交换机端口	设 备	IP 地址	掩 码
10	0/0/1-0/0/5	PC1	192.168.1.101	255.255.255.0
20	0/0/6-0/0/10	PC2	192.168.1.102	255.255.255.0

步骤 2：在交换机上创建 VLAN 并验证，代码如下。

```
switch>enable
Password:                              // 输入特权密码 digitalchina
switch#config terminal                 // 进入全局配置模式
switch(config)#vlan 10                 // 创建 VLAN10
switch(config-vlan10)#exit             // 返回全局模式
switch(config)#vlan 20                 // 创建 VLAN20
switch(config-vlan20)#exit
```

验证配置，代码如下。

```
switch#show vlan
VLAN Name          Type      Media    Ports
---- ------------- --------- -------- ----------------------------------------
1    default       Static    ENET     Ethernet0/0/1      Ethernet0/0/2
                                      Ethernet0/0/3      Ethernet0/0/4
                                      Ethernet0/0/5      Ethernet0/0/6
                                      Ethernet0/0/7      Ethernet0/0/8
                                      Ethernet0/0/9      Ethernet0/0/10
                                      Ethernet0/0/11     Ethernet0/0/12
                                      Ethernet0/0/13     Ethernet0/0/14
                                      Ethernet0/0/15     Ethernet0/0/16
                                      Ethernet0/0/17     Ethernet0/0/18
                                      Ethernet0/0/19     Ethernet0/0/20
                                      Ethernet0/0/21     Ethernet0/0/22
                                      Ethernet0/0/23     Ethernet0/0/24
                                      Ethernet0/0/25     Ethernet0/0/26
10   VLAN0010      Static    ENET
20   VLAN0020      Static    ENET
switch#
```

可以看到已经创建了 VLAN，但 VLAN 中还没有端口。

步骤 3：给 VLAN 添加端口，代码如下。

```
switch(config)#vlan 10
switch(config-vlan10)#switchport interface ethernet 0/0/1-5
Set the port Ethernet0/0/1 access vlan 10 successfully
Set the port Ethernet0/0/2 access vlan 10 successfully
Set the port Ethernet0/0/3 access vlan 10 successfully
Set the port Ethernet0/0/4 access vlan 10 successfully
Set the port Ethernet0/0/5 access vlan 10 successfully
switch(config-vlan10)#exit
```

```
switch(config)#vlan 20
switch(config-vlan20)#switchport interface ethernet 0/0/6-10
Set the port Ethernet0/0/6 access vlan 20 successfully
Set the port Ethernet0/0/7 access vlan 20 successfully
Set the port Ethernet0/0/8 access vlan 20 successfully
Set the port Ethernet0/0/9 access vlan 20 successfully
Set the port Ethernet0/0/10 access vlan 20 successfully
switch(config-vlan20)#exit
switch(config)#
```

验证配置，代码如下。

```
switch#show vlan
VLAN Name          Type      Media    Ports
---- ------------  --------  -------  ----------------------------------
1    default       Static    ENET     Ethernet0/0/11    Ethernet0/0/12
                                      Ethernet0/0/13    Ethernet0/0/14
                                      Ethernet0/0/15    Ethernet0/0/16
                                      Ethernet0/0/17    Ethernet0/0/18
                                      Ethernet0/0/19    Ethernet0/0/20
                                      Ethernet0/0/21    Ethernet0/0/22
                                      Ethernet0/0/23    Ethernet0/0/24
                                      Ethernet0/0/25    Ethernet0/0/26
10   VLAN0010      Static    ENET     Ethernet0/0/1     Ethernet0/0/2
                                      Ethernet0/0/3     Ethernet0/0/4
                                      Ethernet0/0/5
20   VLAN0020      Static    ENET     Ethernet0/0/6     Ethernet0/0/7
                                      Ethernet0/0/8     Ethernet0/0/9
                                      Ethernet0/0/10
switch#
```

步骤 4：验证互通性，见表 2-2。

表 2-2

PC1 位置	PC2 位置	动 作	结 果	备 注
0/0/1	0/0/2	PC1 ping PC2	通	PC1 与 PC2 同在 VLAN10 中
0/0/6	0/0/7	PC1 ping PC2	通	PC1 与 PC2 同在 VLAN20 中
0/0/1	0/0/6	PC1 ping PC2	不通	PC1 与 PC2 不在同一 VLAN 中
0/0/8	0/0/3	PC1 ping PC2	不通	PC1 与 PC2 不在同一 VLAN 中

实训 2　跨交换机 VLAN

VLAN 与 TRUNK

实训目标

小张：师傅，咱们部门在二楼和三楼都有办公室，也是用一台交换机连接的吗？

李工：当然不行。这样做布线很困难，并且一台盒式交换机一般只有 24 或 48 个接口。如果用户数量超过端口限制，则需要增加交换机来实现用户接入。在这种情况下，可以使用跨交换机的 VLAN 来实现多台交换机间用户数据的互联互通。

通过学习本实训，能够掌握跨交换机 VLAN 的概念，具备配置跨交换机 VLAN 能力。

实训拓扑

实训拓扑图如图 2-2 所示。

图　2-2

实训任务

步骤 1：规划交换机的 VLAN 及端口，分配 PC 的 IP 地址，见表 2-3。

表　2-3

VLAN	交换机端口	设备	IP 地址	掩码
10	S1　0/0/1	PC1	192.168.1.10	255.255.255.0
	S2　0/0/10	PC3	192.168.1.11	255.255.255.0
20	S1　0/0/2	PC2	192.168.2.10	255.255.255.0
	S2　0/0/11	PC4	192.168.2.11	255.255.255.0
Trunk 端口	S1　0/0/24	—	—	—
	S2　0/0/24	—	—	—

步骤 2：理解交换机的端口类型。

小张：师傅，什么是 Trunk 端口？

李工：交换机常见的端口类型有 Access 端口和 Trunk 端口。Access 端口主要用来接入终端设备，如 PC、服务器等。这种端口只能承载一个 VLAN 的流量。当它接收到一个来自终端设备的以太网帧时，会打上自己的 PVID（Port Vlan ID）然后转发，如图 2-3 所示。当 Access 端口向终端设备发送数据时，先判断数据帧的 VID 是否和自己的 PVID 一致，如果一致，则剥离 VID 信息转发给终端设备，否则丢弃，如图 2-4 所示。

Trunk 端口主要用于交换机之间或交换机与上层网络设备之间的连接，它可以承载多个 VLAN 的流量。当它接收到一个来自 Access 端口的数据帧时，如果此数据帧的 VID 与自己的 PVID 一致，则剥离 VID 信息后转发到干道链路。如果不一致，则直接转发至干道链路。当它接收到来自干道链路的数据帧时，如果该数据帧无 VID 信息，则打上自己的 PVID 后转发至 Access 端口。如果有 VID 信息，则直接转发至 Access 端口，如图 2-5、图 2-6 所示。

除了 Trunk 和 Access 以外，还有第三种 Hybrid 的端口类型。Hybrid 端口类似于 Trunk 端口，可以传递多个 VLAN。它和 Trunk 端口的区别是，Hybrid 允许传递的 VLAN 中有多个 VLAN 不打 PVID。当 Hybrid 允许多个 VLAN 不打 PVID 时，收到一个没有 PVID 的报文依然使用端口默认的 PVID（Hybrid 端口的默认 PVID 需要手动指定）进行转发。

图 2-3

图 2-4

图 2-5

图 2-6

步骤 3：在交换机 S1 中划分 VLAN、分配端口及 Trunk 口，代码如下。

```
S1(config)#vlan 10
S1(config-vlan10)#switchport interface ethernet 0/0/1
Set the port Ethernet0/0/1 access vlan 10 successfully
S1(config-vlan10)#exit
S1(config)#vlan 20
S1(config-vlan20)#switchport interface ethernet 0/0/2
Set the port Ethernet0/0/2 access vlan 20 successfully
S1(config-vlan20)#exit
S1(config)#interface ethernet 0/0/24
S1(config-if-ethernet0/0/24)#switchport mode trunk
Set the port Ethernet0/0/24 mode Trunk successfully
S1(config-if-ethernet0/0/24)#switchport trunk allowed vlan all
set the Trunk port Ethernet0/0/24 allowed vlan successfully
S1(config-if-ethernet0/0/24)#exit
S1(config)#
```

步骤 4：在交换机 S2 中划分 VLAN、分配端口及 Trunk 端口，代码如下。

```
S2(config)#vlan 10
S2(config-vlan10)#switchport interface ethernet 0/0/10
Set the port Ethernet0/0/10 access vlan 10 successfully
S2(config-vlan10)#exit
S2(config)#vlan 20
S2(config-vlan20)#switchport interface ethernet 0/0/11
Set the port Ethernet0/0/11 access vlan 20 successfully
S2(config-vlan20)#exit
S2(config)#interface ethernet 0/0/24
S2(config-if-ethernet0/0/24)#switchport mode trunk
Set the port Ethernet0/0/24 mode Trunk successfully
S2(config-if-ethernet0/0/24)#switchport trunk allowed vlan all
set the Trunk port Ethernet0/0/24 allowed vlan successfully
S2(config-if-ethernet0/0/24)#exit
S2(config)#
```

步骤 5：验证互通性，见表 2-4。

表 2-4

源主机	端口	目标主机	端口	动作	结果
PC1	S1 0/0/1	PC2	S1 0/0/2	Ping	不通
		PC3	S2 0/0/10	Ping	通
		PC4	S2 0/0/11	Ping	不通

实训 3　生成树协议

RSTP 与 MSTP

实训目标

李工：在交换网络中，为避免因环路而产生广播风暴和交换设备地址系统失效的问题，同时使用冗余链路提高网络的可靠性，交换机提供了生成树协议来解决这些问题。

小张：广播风暴？地址失效？冗余链路？

李工：所谓广播风暴，是指在交换网络中每台交换机不停地从每个端口将同一个数据转发出去，当从某一端口收到时又再一次进行所有端口的转发，这样，交换机的带宽将被耗尽而无法对正常数据进行转发，从而形成广播风暴。

在交换机中，每一台所连接设备的 MAC 地址与所连接的端口号形成的对应表称为 MAC 地址表。伴随着广播风暴的出现，交换机中的 MAC 地址表会出现混乱与错误，加深了广播风暴的影响。

交换网络在物理上形成环路时要在逻辑上阻断环路，形成冗余链路来提高网络的可靠性，这就是生成树协议。

通过学习本实训，能够掌握生成树协议的概念，具备配置 MSTP 功能的能力。

实训拓扑

实训拓扑图如图 2-7 所示。

图　2-7

实训任务

步骤 1：了解广播风暴的危害。

设置 PC1 的 IP 地址为 192.168.1.10，PC2 的 IP 地址为 192.168.1.11，PC1 ping PC2，结果如下。

```
C:\Documents and Settings\Administrator>ping 192.168.1.11
```

```
Pinging 192.168.1.11 with 32 bytes of data:

Request timed out.
Request timed out.
Request timed out.
Request timed out.

Ping statistics for 192.168.1.11:
    Packets: Sent = 4, Received = 4, Lost = 0 (0% loss),
Approximate round trip times in milli-seconds:
    Minimum = 1ms, Maximum = 4ms, Average = 1ms
```

由于生成树协议在神州数码交换机中默认是关闭的，可以看到此时无法 Ping 通，交换机 Link 指示灯频繁闪烁，PC 反应迟缓，整个网络已被广播风暴袭击，无法进行正常通信。

步骤 2：理解生成树协议。

生成树协议要解决的问题是，在网络中有物理环路的情况下，从逻辑上阻断环路，避免广播风暴的发生，同时，在链路出现故障时，能够自动启用冗余链路恢复通信。

IEEE 802.1d 协议通过在交换机上运行一套复杂的算法 STA（Spanning-Tree Algorithm），使冗余端口置于"阻断状态"，使得接入网络的计算机在与其他计算机通信时，只有一条链路生效，而当这个链路出现故障无法使用时，IEEE 802.1d 协议会重新计算网络链路，将处于"阻断状态"的端口重新打开，从而既保障了网络正常运转，又保证了冗余能力。

生成树协议通过选择根网桥、根端口、指定端口 3 个步骤来完成生成树的计算。

步骤 3：选择根网桥。

根网桥的选择是通过交换机的网桥 ID 来确定的。打开交换机的生成树协议，并查询交换机的网桥 ID，代码如下。

```
S1(config)#spanning-tree                              // 开启生成树协议

MSTP is starting now, please wait.............
MSTP is enabled successfully.
S1(config)#exit
S1#show spanning-tree

                -- MSTP Bridge Config Info --

Standard        :  IEEE 802.1s
Bridge MAC      :  00:03:0f:0f:2a:63
Bridge Times :  Max Age 20, Hello Time 2, Forward Delay 15
Force Version:  3
```

```
########################## Instance 0 ##########################
Self Bridge Id     : 32768 -   00:03:0f:0f:2a:63
Root Id            : this switch
Ext.RootPathCost : 0
Region Root Id     : this switch
Int.RootPathCost : 0
Root Port ID       : 0
Current port list in Instance 0:
Ethernet0/0/1 Ethernet0/0/2 (Total 2)


PortName        ID        ExtRPC    IntRPC   State Role     DsgBridge         DsgPort

--------------  --------  --------  -------- ---   ----  ----------------  -------

 Ethernet0/0/1 128.001      0          0   FWD  DSGN  32768.00030f0f2a63 128.001
 Ethernet0/0/2 128.002      0          0   FWD  DSGN  32768.00030f0f2a63 128.002
S1#
```

其中，Self Bridge Id ：32768 - 00:03:0f:0f:2a:63 就是该交换机的网桥 ID。它由两个部分组成，前面的数字 32768 是交换机的默认优先级，其取值范围为 0 ～ 65535，步长为 4096，数字越小则优先级越高；后面是该交换机的 MAC 地址。若各交换机优先级相同，则比较交换机的 MAC 地址，越小则优先级越高。

同样可以查询到 S2 的网桥 ID 为 Self Bridge Id ：32768 - 00:03:0f:0f:2a:7d，交换机 S3 的网桥 ID 为 Self Bridge Id ：32768 - 00:03:0f:0f:1c:8e。

比较三个网桥 ID 的大小，它们的优先级相同，而 MAC 地址中最小的是交换机 S3。因此，S3 被生成树协议选定为根网桥。

步骤 4：选择根端口。

根网桥确定后，需要在非根网桥上选择根端口。选择根端口的依据是：

1）端口到达根网桥的路径成本最低。路径成本是指从非根网桥到根网桥上所有链路的成本之和。神州数码设备默认 10Mbit/s/100Mbit/s 自适应的路径开销为 200 000。

2）若路径成本无法选择根端口，则依据非根网桥的直连（上游）网桥的桥 ID 最小来选择根端口，在交换机级联时可能出现此种情况。

3）若依然无法确定根端口，则再依据上游端口 ID 最小来确定根端口。端口 ID 如 128.001，前面的 128 为端口优先级，其取值范围为 0 ～ 255，步长为 1，默认值为 128。后面的 001 为端口编号。

根据以上依据，在此例中，S3 为根网桥，需在非根网桥 S1 和 S2 中选择根端口。

S1 端口 E0/0/1 路径成本 =200 000+200 000=400 000；

S1 端口 E0/0/2 路径成本 =200 000；

S2 端口 E0/0/1 路径成本 =200 000；
S2 端口 E/0/0/2 路径成本 =200 000+200 000=400 000；
所以，S1 的 E/0/0/2 和 S2 的 E/0/0/1 端口为根端口。

步骤 5：选择指定端口。

每条连接交换机的链路上都需要选择指定端口。根网桥的所有端口都是指定端口，非根网桥的指定端口依据以下方法选择：

1）根路径成本最低。
2）端口所在网桥的桥 ID 值较小。
3）端口 ID 值较小。

在此例中，S3 为根网桥，其所有端口都是指定端口，S1 的 E0/0/2 和 S2 的 E0/0/1 为根端口，所以，需要在 S1 的 E0/0/1 和 S2 的 E0/0/2 中选择指定端口。

先比较根路径成本如下。

S1 的 E0/0/1 到根网桥的路径成本：经过 S2 交换机连接到 S3，故路径成本为 400 000，同样，S2 的 E/0/0/2 根路径成本也是 400 000，无法比较。

再比较端口所在网桥的桥 ID 值。如前所述，S1 的桥 ID 为 32768 - 00:03:0f:0f:2a:63，S2 的桥 ID 为 32768 - 00:03:0f:0f:2a:7d，S1 的桥 ID 值较小，所以 S1 的 E0/0/1 端口为指定端口。

最后，将既非根端口又非指定端口的 S2 E0/0/2 端口关闭，生成树协议计算完成，封闭了物理环路，并将 S1 的 E0/0/1 到 S2 的 E0/0/2 链路作为冗余链路。

步骤 6：验证生成树的冗余性。

李工：小张，你来验证一下。

小张：好的。我用 PC1 ping PC2，代码如下。

```
C:\Documents and Settings\Administrator>ping 192.168.1.11

Pinging 192.168.1.11 with 32 bytes of data:

Reply from 192.168.1.11: bytes=32 time=4ms TTL=64
Reply from 192.168.1.11: bytes=32 time=1ms TTL=64
Reply from 192.168.1.11: bytes=32 time=1ms TTL=64
Reply from 192.168.1.11: bytes=32 time=1ms TTL=64

Ping statistics for 192.168.1.11:
    Packets: Sent = 4, Received = 4, Lost = 0 (0% loss),
Approximate round trip times in milli-seconds:
    Minimum = 1ms, Maximum = 4ms, Average = 1ms
```

可以正常通信，没有广播风暴发生，生成树协议工作正常。

查看 S2 的生成树协议信息，代码如下。

```
S2#show spanning-tree
```

```
                    -- MSTP Bridge Config Info --

Standard        :  IEEE 802.1s
Bridge MAC      :  00:03:0f:0f:2a:7d
Bridge Times :    Max Age 20, Hello Time 2, Forward Delay 15
Force Version:   3


########################### Instance 0 ###########################
Self Bridge Id      : 32768 -   00:03:0f:0f:2a:7d
Root Id             : 32768 -   00:03:0f:0f:1c:8e
Ext.RootPathCost : 200000
Region Root Id      : this switch
Int.RootPathCost : 0
Root Port ID        : 128.1
Current port list in Instance 0:
Ethernet0/0/1 Ethernet0/0/2 (Total 2)


PortName        ID       ExtRPC    IntRPC   State Role     DsgBridge            DsgPort

-------------- -------- --------- --------- ---  ---- ----------------- -------

 Ethernet0/0/1 128.001         0          0 FWD  DSGN 32768.00030f0f2a7d 128.001
 Ethernet0/0/2 128.002         0          0 BLK  BKUP 32768.00030f0f2a7d 128.002
```

看到 S2 的 E0/0/2 端口处于阻塞状态。

把 S1 到 S3 的链路拔除，再用 PC1 ping PC2，代码如下。

```
C:\Documents and Settings\Administrator>ping 192.168.1.11

Pinging 192.168.1.11 with 32 bytes of data:

Reply from 192.168.1.11: bytes=32 time=2ms TTL=64
Reply from 192.168.1.11: bytes=32 time=1ms TTL=64
Reply from 192.168.1.11: bytes=32 time=12ms TTL=64
Reply from 192.168.1.11: bytes=32 time=1ms TTL=64

Ping statistics for 192.168.1.11:
    Packets: Sent = 4, Received = 4, Lost = 0 (0% loss),
Approximate round trip times in milli-seconds:
```

Minimum = 1ms, Maximum = 12ms, Average = 4ms

仍然可以正常通信,再次查看 S2 的生成树协议,代码如下。

```
S2#show spanning-tree

                -- MSTP Bridge Config Info --

Standard      :   IEEE 802.1s
Bridge MAC    :   00:03:0f:0f:2a:7d
Bridge Times  :   Max Age 20, Hello Time 2, Forward Delay 15
Force Version:    3


########################### Instance 0 ###########################
Self Bridge Id      : 32768 -   00:03:0f:0f:2a:7d
Root Id             : 32768 -   00:03:0f:0f:1c:8e
Ext.RootPathCost : 200000
Region Root Id      : this switch
Int.RootPathCost : 0
Root Port ID        : 128.1
Current port list in Instance 0:
Ethernet0/0/1 Ethernet0/0/2 (Total 2)


PortName         ID      ExtRPC    IntRPC   State Role    DsgBridge          DsgPort

--------------- -------- --------- --------- ---  ----  ------------------ -------

 Ethernet0/0/1 128.001         0          0 FWD  DSGN  32768.00030f0f2a7d 128.001
 Ethernet0/0/2 128.002         0          0 FWD  DSGN  32768.00030f0f2a7d 128.002
```

看到 E0/0/2 端口已处于转发状态,冗余链路工作正常。

步骤 7:修改优先级改变生成树形态。

李工:前面使用了交换机及端口的默认优先级来运行生成树协议。在有些情况下,需要特别指定某台交换机为根网桥,或特别指定某端口为指定端口时,可以采用改变交换机或端口的优先级来达到目的。

在本例中,如果指定 S1 为根网桥,则需要以 4096 为倍数减小 S1 的优先级值,使 S1 获得最高优先权。

```
S1(Config)#spanning-tree mst 0 priority 28672              // 改变 S1 交换机优先级
```

查看 S2 交换机生成树协议,代码如下。

```
S2#show spanning-tree
```

```
                -- MSTP Bridge Config Info --

Standard        :   IEEE 802.1s
Bridge MAC      :   00:03:0f:0f:2a:7d
Bridge Times :   Max Age 20, Hello Time 2, Forward Delay 15
Force Version:   3

########################## Instance 0 ##########################
Self Bridge Id     : 32768 -    00:03:0f:0f:2a:7d
Root Id            : 28672 -    00:03:0f:0f:2a:63
Ext.RootPathCost : 200000
Region Root Id     : this switch
Int.RootPathCost : 0
Root Port ID       : 128.2
Current port list in Instance 0:
Ethernet0/0/1 Ethernet0/0/2 (Total 2)


PortName         ID       ExtRPC    IntRPC    State Role    DsgBridge           DsgPort

--------------- --------- --------- --------- ---  ----  -----------------  -------

Ethernet0/0/1 128.001     0         0 BLK  BKUP 32768.00030f0f2a7d 128.001
Ethernet0/0/2 128.002     0         0 FWD  DSGN 32768.00030f0f2a7d 128.002
```

可以看到 S2 的端口 E0/0/1 被阻塞，E0/0/2 成为根端口，生成树形态已被改变。

实训 4　多实例生成树协议

RSTP 与 MSTP

实训目标

李工：相对于基本生成树，多实例生成树允许多个具有相同拓扑的 VLAN 映射到一个生成树实例上，而这个生成树拓扑同其他生成树实例相互独立。这种机制多重生成树实例为映射到它的 VLAN 的数据流量提供了独立的发送路径，实现不同实例间 VLAN 数据流量的负载分担。

多实例生成树由于多个 VLAN 可以映射到一个单一的生成树实例，IEEE 802.1s 委员会提出了 MST 域的概念，用来解决如何判断某个 VLAN 映射到哪个生成树实例的问题。在这个实训环境中可以进一步理解多 VLAN 的生成树协议原理和实际拓扑生成。

通过学习本实训，能够掌握多实例生成树协议的概念，具备配置多实例 MSTP 功能的能力。

实训拓扑

实训拓扑图如图 2-8 所示。

图 2-8

实训任务

步骤1：规划交换机 VLAN 及端口分配，PC 的 IP 地址，见表 2-5。

表 2-5

VLAN	交换机端口	设 备	IP 地址	掩 码
10	S1　0/0/1	PC1	192.168.1.10	255.255.255.0
	S2　0/0/10	PC3	192.168.1.11	255.255.255.0
20	S1　0/0/2	PC2	192.168.2.10	255.255.255.0
	S2　0/0/11	PC4	192.168.2.11	255.255.255.0
Trunk 端口	S1　0/0/23-24	—	—	—
	S2　0/0/23-24	—	—	—

步骤2：在交换机上划分 VLAN、分配端口及 Trunk 端口。

交换机 S1：

S1(config)#vlan 10

S1(config-vlan10)#switchport interface ethernet 0/0/1

Set the port Ethernet0/0/1 access vlan 10 successfully

S1(config-vlan10)#exit

S1(config)#vlan 20

S1(config-vlan20)#switchport interface ethernet 0/0/2

Set the port Ethernet0/0/2 access vlan 20 successfully

S1(config-vlan20)#exit

S1(config)#interface ethernet 0/0/23-24

S1(config-if-port-range)#switchport mode trunk

Set the port Ethernet0/0/23 mode Trunk successfully

Set the port Ethernet0/0/24 mode Trunk successfully

S1(config-if-port-range)#exit

S1(config)#

交换机 S2：

S2(config)#vlan 10

S2(config-vlan10)#switchport interface ethernet 0/0/10

第 2 章 交换机典型园区应用实践

```
Set the port Ethernet0/0/10 access vlan 10 successfully
S2(config-vlan10)#exit
S2(config)#vlan 20
S2(config-vlan20)#switchport interface ethernet 0/0/11
Set the port Ethernet0/0/11 access vlan 20 successfully
S2(config-vlan20)#exit
S2(config)#interface ethernet 0/0/23-24
S2(config-if-port-range)#switchport mode trunk
Set the port Ethernet0/0/23 mode Trunk successfully
Set the port Ethernet0/0/24 mode Trunk successfully
S2(config-if-port-range)#exit
S2(config)#
```

步骤 3：配置多实例生成树，在交换机 S1、S2 上分别将 VLAN 10 映射到实例 1 上，将 VLAN 20 映射到实例 2 上。

交换机 S1：

```
S1(config)#spanning-tree mst configuration
S1(config-mstp-region)#name mstp
S1(config-mstp-region)#instance 1 vlan 10
S1(config-mstp-region)#instance 2 vlan 20
S1(config-mstp-region)#exit
S1(config)#spanning-tree

MSTP is starting now, please wait.............
MSTP is enabled successfully.
S1(config)#
```

交换机 S2：

```
S2(config)#spanning-tree mst configuration
S2(Config-Mstp-Region)#name mstp
S2(Config-Mstp-Region)#instance 1 vlan 10
S2(Config-Mstp-Region)#instance 2 vlan 20
S2(Config-Mstp-Region)#exit
S2(config)#spanning-tree

MSTP is starting now, please wait...............
MSTP is enabled successfully.
S2(config)#
```

步骤 4：在根交换机中配置端口在不同实例中的优先级，确保不同实例阻塞不同端口，代码如下。

```
S1#show spanning-tree
```

```
                    -- MSTP Bridge Config Info --

Standard         :  IEEE 802.1s
Bridge MAC       :  00:03:0f:0f:2a:63
Bridge Times :    Max Age 20, Hello Time 2, Forward Delay 15
Force Version:    3


########################## Instance 0 ##########################
Self Bridge Id      : 32768 -   00:03:0f:0f:2a:63
Root Id             : this switch
Ext.RootPathCost : 0
Region Root Id      : this switch
Int.RootPathCost : 0
Root Port ID        : 0
............
```

可以看出 S1 交换机为根交换机,在根交换机上修改 Trunk 端口在不同实例中的优先级,代码如下。

```
S1(config)#interface ethernet 0/0/23
S1(config-if-ethernet0/0/23)#spanning-tree mst 1 port-priority 32
S1(config-if-ethernet0/0/23)#exit
S1(config)#interface ethernet 0/0/24
S1(config-if-ethernet0/0/24)#spanning-tree mst 2 port-priority 32
S1(config-if-ethernet0/0/24)#exit
S1(config)#
```

步骤 5:验证多实例生成树,代码如下。

```
S2#show spanning-tree
                    -- MSTP Bridge Config Info --

Standard         :  IEEE 802.1s
Bridge MAC       :  00:03:0f:0f:4a:29
Bridge Times :    Max Age 20, Hello Time 2, Forward Delay 15
Force Version:    3


########################## Instance 0 ##########################
Self Bridge Id      : 32768 -   00:03:0f:0f:4a:29
Root Id             : 32768.00:03:0f:0f:2a:63
Ext.RootPathCost : 0
Region Root Id      : 32768.00:03:0f:0f:2a:63
```

```
Int.RootPathCost : 200000
Root Port ID     : 128.23
Current port list in Instance 0:
Ethernet1/10 Ethernet1/11 Ethernet1/23 Ethernet1/24 (Total 4)

  PortName         ID      ExtRPC    IntRPC  State Role     DsgBridge          DsgPort
 --------------  -------  --------- --------- --- ----  -----------------  -------
  Ethernet1/10  128.010      0       200000  FWD DSGN  32768.00030f0f4a29  128.010
  Ethernet1/11  128.011      0       200000  FWD DSGN  32768.00030f0f4a29  128.011
  Ethernet1/23  128.023      0            0  FWD ROOT  32768.00030f0f2a63  128.023
  Ethernet1/24  128.024      0            0  BLK ALTR  32768.00030f0f2a63  128.024


########################### Instance 1 ###########################
Self Bridge Id   : 32768.00:03:0f:0f:4a:29
Region Root Id   : 32768.00:03:0f:0f:2a:63
Int.RootPathCost : 200000
Root Port ID     : 128.23
Current port list in Instance 1:
Ethernet1/10 Ethernet1/23 Ethernet1/24 (Total 3)

  PortName         ID      IntRPC    State Role     DsgBridge          DsgPort
 --------------  -------  --------- --- ----  -----------------  -------
  Ethernet1/10  128.010   200000    FWD DSGN  32768.00030f0f4a29  128.010
  Ethernet1/23  128.023        0    FWD ROOT  32768.00030f0f2a63  032.023
  Ethernet1/24  128.024        0    BLK ALTR  32768.00030f0f2a63  128.024


########################### Instance 2 ###########################
Self Bridge Id   : 32768.00:03:0f:0f:4a:29
Region Root Id   : 32768.00:03:0f:0f:2a:63
Int.RootPathCost : 200000
Root Port ID     : 128.24
Current port list in Instance 2:
Ethernet1/11 Ethernet1/23 Ethernet1/24 (Total 3)

  PortName         ID      IntRPC    State Role     DsgBridge          DsgPort
 --------------  -------  --------- --- ----  -----------------  -------
  Ethernet1/11  128.011   200000    FWD DSGN  32768.00030f0f4a29  128.011
  Ethernet1/23  128.023        0    BLK ALTR  32768.00030f0f2a63  128.023
  Ethernet1/24  128.024        0    FWD ROOT  32768.00030f0f2a63  032.024
S2#
```

可以看到在不同实例中 23、24 端口的转发阻塞情况。

实训 5　链路聚合

链路聚合

实训目标

小张：这个链路聚合看上去和生成树有点像。

李工：虽然有点形似，但功能完全不同。链路聚合是将几个链路作聚合处理，这几个链路必须是同时连接两个相同的设备，这样链路聚合就可以实现几个链路相加的带宽。比如，可以将 4 个 100Mbit/s 链路使用链路聚合做成一个逻辑链路，这样在全双工条件下就可以达到 800Mbit/s 的带宽，同时可以实现链路负载平衡，提供额外的容错性能。这种方式比较经济，实现也相对容易。

实现链路聚合的条件：

① 端口必须处于相同的 VLAN 之中。

② 端口必须使用相同的传输介质。

③ 端口必须都处于全双工工作模式。

④ 端口必须是相同传输速率的端口。

通过学习本实训，能够掌握链路聚合的概念，具备配置 PORT-GROUP 功能的能力。

实训拓扑

实训拓扑图如图 2-9 所示。

图　2-9

实训任务

步骤 1：正确连接并创建端口组。

由于神州数码交换机默认生成树协议是关闭的，所以在创建链路聚合时只能先连接一条线，避免产生广播风暴，待聚合端口创建成功后再连接其他线缆。

交换机 S1：

S1(Config)#port-group 1

S1(Config)#

交换机 S2：
```
S2(Config)#port-group 2
S2(Config)#
```
步骤 2：静态生成链路聚合通道。
交换机 S1：
```
S1(config)#interface ethernet 0/0/1-2
S1(config-if-port-range)#port-group 1 mode on
S1(config-if-port-range)#
```
交换机 S2：
```
S2(config)#interface ethernet 0/0/1-2
S2(config-if-port-range)#port-group 2 mode on        //注意与交换机 S1 的模式匹配
S2(config-if-port-range)#
```
查看 VLAN 信息：

```
S1#show vlan
VLAN Name          Type      Media     Ports
---- ------------- --------- --------- --------------------------------
1    default       Static    ENET      Ethernet0/0/3      Ethernet0/0/4
                                       Ethernet0/0/5      Ethernet0/0/6
                                       Ethernet0/0/7      Ethernet0/0/8
                                       Ethernet0/0/9      Ethernet0/0/10
                                       Ethernet0/0/11     Ethernet0/0/12
                                       Ethernet0/0/13     Ethernet0/0/14
                                       Ethernet0/0/15     Ethernet0/0/16
                                       Ethernet0/0/17     Ethernet0/0/18
                                       Ethernet0/0/19     Ethernet0/0/20
                                       Ethernet0/0/21     Ethernet0/0/22
                                       Ethernet0/0/23     Ethernet0/0/24
                                       Ethernet0/0/25     Ethernet0/0/26
                                       Port-Channel1
S1#
```

可以看到多了一个通道 Port-Channel1（在 S2 上显示为 Port-Channel2），表示此时静态聚合链路已创建完成，分别连接 S1 和 S2 交换机的 E0/0/1 和 E0/0/2 端口，使用 PC1 ping PC2，代码如下。

```
C:\Documents and Settings\Administrator>ping 192.168.1.11

Pinging 192.168.1.11 with 32 bytes of data:

Reply from 192.168.1.11: bytes=32 time=2ms TTL=64
```

```
Reply from 192.168.1.11: bytes=32 time=1ms TTL=64
Reply from 192.168.1.11: bytes=32 time=1ms TTL=64
Reply from 192.168.1.11: bytes=32 time=1ms TTL=64


Ping statistics for 192.168.1.11:
    Packets: Sent = 4, Received = 4, Lost = 0 (0% loss),
Approximate round trip times in milli-seconds:
    Minimum = 1ms, Maximum = 12ms, Average = 4ms
```

可以正常通信。拔掉聚合链路中的一条链路,再用 PC1 ping PC2,代码如下。

```
C:\Documents and Settings\Administrator>ping 192.168.1.11

Pinging 192.168.1.11 with 32 bytes of data:

Reply from 192.168.1.11: bytes=32 time=2ms TTL=64
Reply from 192.168.1.11: bytes=32 time=2ms TTL=64
Reply from 192.168.1.11: bytes=32 time=2ms TTL=64
Reply from 192.168.1.11: bytes=32 time=1ms TTL=64


Ping statistics for 192.168.1.11:
    Packets: Sent = 4, Received = 4, Lost = 0 (0% loss),
Approximate round trip times in milli-seconds:
    Minimum = 1ms, Maximum = 12ms, Average = 4ms
```

依然通信正常,静态链路聚合创建完成。

步骤 3:LACP 动态生成链路聚合通道。

启动 LACP 的端口可以有两种工作模式:Passive 和 Active。

Passive:被动模式,该模式下端口不会主动发送 LACPDU 报文,在接收到对端发送的 LACP 报文后,该端口进入协议计算状态。

Active:主动模式,该模式下端口会主动向对端发送 LACPDU 报文,进行 LACP 的计算。

交换机 S1:

```
S1(Config)#port-group 1
S1(config)#interface ethernet 0/0/1-2
S1(config-if-port-range)#port-group 1 mode active
S1(config-if-port-range)#
```

交换机 S2:

```
S2(Config)#port-group 2
S2(config)#interface ethernet 0/0/1-2
S2(config-if-port-range)#port-group 2 mode passive
S2(config-if-port-range)#
```

验证 S1：

```
S1#show interface port-channel 1
Interface brief:
    Port-Channel1 is up, line protocol is up
    Port-Channel1 is layer 2 port, alias name is (null), index is 29
    Port-Channel1 is LAG port, member is :
        Ethernet0/0/1        Ethernet0/0/3
    Hardware is EtherChannel, address is 00-03-0f-1a-ec-3d
```

可以看到 S1 上的聚合组 Port-Channel1 已经 up，聚合端口有 E0/0/1 和 E0/0/3 两个端口。

验证 S2：

```
S2# show interface port-channel 2
Interface brief:
    Port-Channel2 is up, line protocol is up
    Port-Channel2 is layer 2 port, alias name is (null), index is 30
    Port-Channel2 is LAG port, member is :
        Ethernet0/0/1        Ethernet0/0/3
    Hardware is EtherChannel, address is 00-03-0f-1a-ec-49
```

S2 上的聚合组 Port-Channel2 也已经 up。

PC1 ping PC2：

```
C:\Documents and Settings\Administrator>ping 192.168.1.11

Pinging 192.168.1.11 with 32 bytes of data:

Reply from 192.168.1.11: bytes=32 time=2ms TTL=64
Reply from 192.168.1.11: bytes=32 time=1ms TTL=64
Reply from 192.168.1.11: bytes=32 time=1ms TTL=64
Reply from 192.168.1.11: bytes=32 time=1ms TTL=64

Ping statistics for 192.168.1.11:
    Packets: Sent = 4, Received = 4, Lost = 0 (0% loss),
Approximate round trip times in milli-seconds:
    Minimum = 1ms, Maximum = 12ms, Average = 4ms
```

动态链路聚合完成。

实训 6　端口安全

二层安全技术

实训目标

李工：有时为了安全和管理上的需要，需将设备或终端的 MAC 地址与交换机端口进行绑定，端口只允许已绑定 MAC 地址的数据流的转发。即，MAC 地址与端口绑定后，该 MAC

地址的数据流只能从绑定端口进入，其他没有与端口绑定的 MAC 地址的数据流不可以从该端口进入。

通过学习本实训，能够掌握端口安全的概念，具备配置 MAC 地址绑定功能的能力。

实训拓扑

实训拓扑图如图 2-10 所示。

图 2-10

实训任务

步骤 1：规划 PC 的 IP 地址与端口，见表 2-6。

表 2-6

设备	IP 地址	掩码	MAC 地址	端口
PC1	192.168.1.10	255.255.255.0	00-0b-4c-9e-2a-1c	E0/0/1
PC2	192.168.1.11	255.255.255.0	00-0b-2e-5d-62-4e	E0/0/2
PC3	192.168.1.12	255.255.255.0	00-0b-cd-4a-97-08	E0/0/3

步骤 2：静态绑定 MAC 地址。

E0/0/1 端口绑定 PC1 的 MAC 地址，代码如下。

```
S1(config-if-ethernet0/0/1)#switchport port-security
S1(config-if-ethernet0/0/1)#switchport port-security mac-address 00-0b-4c-9e-2a-1c
```

PC1 分别 Ping PC2 和 PC3 可以 ping 通，代码如下。

```
C:\Documents and Settings\Administrator>ping 192.168.1.12

Pinging 192.168.1.12 with 32 bytes of data:

Reply from 192.168.1.12: bytes=32 time<1ms TTL=64
Reply from 192.168.1.12: bytes=32 time<1ms TTL=64
Reply from 192.168.1.12: bytes=32 time<1ms TTL=64
Reply from 192.168.1.12: bytes=32 time<1ms TTL=64

Ping statistics for 192.168.1.12:
    Packets: Sent = 4, Received = 4, Lost = 0 (0% loss),
```

Approximate round trip times in milli-seconds:

　　Minimum = 0ms, Maximum = 0ms, Average = 0ms

把 PC1 更换到 E0/0/4 端口，再次 ping PC2 和 PC3，已无法 ping 通，代码如下。

C:\Documents and Settings\Administrator>ping 192.168.1.12

Pinging 192.168.1.12 with 32 bytes of data:

Request timed out.
Request timed out.
Request timed out.
Request timed out.

Ping statistics for 192.168.1.12:

　　Packets: Sent = 4, Received = 0, Lost = 4 (100% loss),

可见，PC1 只能通过其绑定的 E0/0/1 端口与网络进行通信。

小张：如果把其他 PC 换到 E0/0/1 端口是什么情况呢？

李工：你试一下。

小张把 PC2 更换到 E0/0/1 端口，分别是 ping E0/0/3 端口的 PC3 和 E0/0/4 端口的 PC1，代码如下。

C:\Documents and Settings\Administrator>ping 192.168.1.12

Pinging 192.168.1.12 with 32 bytes of data:

Reply from 192.168.1.12: bytes=32 time<1ms TTL=64
Reply from 192.168.1.12: bytes=32 time<1ms TTL=64
Reply from 192.168.1.12: bytes=32 time<1ms TTL=64
Reply from 192.168.1.12: bytes=32 time<1ms TTL=64

Ping statistics for 192.168.1.12:

　　Packets: Sent = 4, Received = 4, Lost = 0 (0% loss),

Approximate round trip times in milli-seconds:

　　Minimum = 0ms, Maximum = 0ms, Average = 0ms

C:\Documents and Settings\Administrator>ping 192.168.1.10

Pinging 192.168.1.10 with 32 bytes of data:

Request timed out.

```
Request timed out.
Request timed out.
Request timed out.

Ping statistics for 192.168.1.10:
    Packets: Sent = 4, Received = 0, Lost = 4 (100% loss),
```

小张：为什么 PC2 还能 ping 通 PC3 呢？

李工：端口 E0/0/1 绑定了 PC1，依然可以与其他 MAC 地址通信，但 PC1 只能通过 E0/0/1 端口进行通信。如果在 E0/0/1 口加入指令：

```
S1(config-if-ethernet0/0/1)#switchport port-security lock
```

再用 PC2 ping PC3：

```
C:\Documents and Settings\Administrator>ping 192.168.1.12

Pinging 192.168.1.12 with 32 bytes of data:

Request timed out.
Request timed out.
Request timed out.
Request timed out.

Ping statistics for 192.168.1.12:
    Packets: Sent = 4, Received = 0, Lost = 4 (100% loss),
```

就无法 ping 通了。

把 PC1 换回 E0/0/1 端口，ping PC3：

```
C:\Documents and Settings\Administrator>ping 192.168.1.12

Pinging 192.168.1.12 with 32 bytes of data:

Reply from 192.168.1.12: bytes=32 time<1ms TTL=64
Reply from 192.168.1.12: bytes=32 time<1ms TTL=64
Reply from 192.168.1.12: bytes=32 time<1ms TTL=64
Reply from 192.168.1.12: bytes=32 time<1ms TTL=64

Ping statistics for 192.168.1.12:
    Packets: Sent = 4, Received = 4, Lost = 0 (0% loss),
Approximate round trip times in milli-seconds:
    Minimum = 0ms, Maximum = 0ms, Average = 0ms
```

可以看到，此时 E0/0/1 端口只能接受绑定的 MAC 地址，而不能接受其他 MAC 地址通过此端口进行通信。

小张：那是不是一个端口只能静态绑定一个 MAC 地址呢？

李工：当然不是。在默认情况下，交换机允许绑定一个 MAC 地址，但可以让一个端口绑定多个 MAC 地址。

步骤 3：静态绑定多个 MAC 地址。

取消端口 E0/0/1 锁定，更改最大 MAC 地址数，然后绑定 PC3 的 MAC 地址，代码如下。

```
S1(config-if-ethernet0/0/1)#no switchport port-security lock
S1(config-if-ethernet0/0/1)#switchport port-security maximum 2
S1(config-if-ethernet0/0/1)#switchport port-security mac-address 00-0b-cd-4a-97-08
S1(config-if-ethernet0/0/1)#switchport port-security lock
```

查看绑定的 MAC 地址情况，代码如下。

```
S1#show port-security address
Security Mac Address Table
------------------------------------------------------------------------
Vlan    Mac Address             Type                Ports
 1      00-0b-4c-9e-2a-1c       SecurityConfigured  Ethernet0/0/1
 1      00-0b-cd-4a-97-08       SecurityConfigured  Ethernet0/0/1
------------------------------------------------------------------------
Total Addresses in System :2
Max Addresses limit in System :128

S1#
```

可以看出，E0/0/1 端口已经绑定了 2 个 MAC 地址，最多可以绑定 128 个 MAC 地址。

把 PC3 插在 E0/0/1 端口，PC1 插在 E0/0/3 端口，PC2 端口不变，用 PC3 ping PC2，代码如下。

```
C:\Documents and Settings\Administrator>ping 192.168.1.11

Pinging 192.168.1.12 with 32 bytes of data:

Reply from 192.168.1.12: bytes=32 time=1ms TTL=64
Reply from 192.168.1.12: bytes=32 time<1ms TTL=64
Reply from 192.168.1.12: bytes=32 time<1ms TTL=64
Reply from 192.168.1.12: bytes=32 time<1ms TTL=64

Ping statistics for 192.168.1.12:
    Packets: Sent = 4, Received = 4, Lost = 0 (0% loss),
Approximate round trip times in milli-seconds:
    Minimum = 0ms, Maximum = 0ms, Average = 0ms
```

可以 ping 通。PC3 如果现在 ping PC1，代码如下。

```
C:\Documents and Settings\Administrator>ping 192.168.1.10

Pinging 192.168.1.10 with 32 bytes of data:

Request timed out.
Request timed out.
Request timed out.
Request timed out.

Ping statistics for 192.168.1.10:
    Packets: Sent = 4, Received = 0, Lost = 4 (100% loss),
```

这是因为 PC1 已绑定在 E0/0/1 端口，不能在其他端口通信。

步骤 4：动态绑定 MAC 地址。

在通常情况下，并不需要特别指定是哪台设备必须绑定在哪个端口。这时，可以使用动态绑定办法，让交换机端口自动绑定连接它的设备。

还原 PC 与交换机端口的连接，重启交换机，还原到出厂设置，代码如下。

```
S1#set default
Are you sure? [Y/N] = y
S1#reload
Process with reboot? [Y/N] y
```

设置 E0/0/1 端口自动绑定连接它的 MAC 地址，代码如下。

```
S1(config-if-ethernet0/0/1)#switchport port-security
S1(config-if-ethernet0/0/1)#switchport port-security lock
S1(config-if-ethernet0/0/1)#switchport port-security convert
1 dynamic mac have been converted to security mac on interface Ethernet0/0/1
S1(config-if-ethernet0/0/1)#
```

查看端口绑定地址，代码如下。

```
S1#show port-security address
Security Mac Address Table
-----------------------------------------------------------------
Vlan    Mac Address             Type                Ports
 1      00-0b-4c-9e-2a-1c       SecurityConfigured  Ethernet0/0/1
-----------------------------------------------------------------
Total Addresses in System :1
Max Addresses limit in System :128

S1#
```

步骤 5：绑定 MAC-IP 地址。

MAC-IP 地址绑定可以将端口绑定到指定的设备和 IP 地址，更换设备或 IP 地址均无法通过此端口访问，代码如下。

```
S1(config)#am enable
S1(config)#interface ethernet1/1
S1(config-if-ethernet1/1)#am port
S1(config-if-ethernet1/1)#am mac-ip-pool 00-0b-4c-9e-2a-1c 192.168.1.10
```

可以将端口 E1/1 绑定为 PC1，此时 PC1 可通过该端口访问 PC2 或 PC3。若更改 PC1 的 IP 地址，或将其他设备（如 PC2）连接 E1/1 端口则无法进行访问。

需要注意的是，PC1 仍可通过未绑定地址的端口访问其他设备。

步骤 6：绑定 IP 地址区间。

将端口绑定一个 IP 地址区间，则此区间的 IP 地址设备可通过该端口进行访问，不在此区间的 IP 地址无法通过该端口访问网络。

将端口 E1/2 绑定 192.168.1.10 和 192.168.1.11 两个 IP 地址，代码如下。

```
S1(config)#am enable
S1(config)#interface ethernet 1/2
S1(config-if-ethernet1/2)#am port
S1(config-if-ethernet1/2)#am ip-pool 192.168.1.10 2    // 绑定 192.168.1.10 开始的 2 个地址
```

除 192.168.1.10 和 192.168.1.11 外的其他 IP 地址均无法通过此端口通信。

第 3 章 交换机路由应用实践

实训 1 单臂路由实现 VLAN 间互访

实训目标

路由技术与应用

李工：划分 VLAN 可以隔离广播域，有效提高网络性能。但 VLAN 之间无法通信，这并不是我们的目的。要既能隔离广播域、防止广播风暴的发生，又能实现 VLAN 之间的通信，就需要网络层的设备来支持。常见的网络层设备是路由器，可以通过路由器以单臂路由方式来实现 VLAN 之间的通信。

通过学习本实训，能够掌握通过单臂路由实现 VLAN 间互访，具备配置单臂路由的能力。

实训拓扑

实训拓扑图如图 3-1 所示。

图 3-1

实训任务

步骤 1：规划交换机端口、VLAN 及 IP 地址等，见表 3-1。

表 3-1

交换机端口	所属 VLAN	连接设备	IP 地址	网 关
E1/1	VLAN10	PC1	192.168.1.10	192.168.1.1
E1/2	VLAN20	PC2	192.168.2.10	192.168.2.1
E1/24	Trunk 端口，连接路由器 F0/0 以太网口			

步骤 2：配置交换机，代码如下。

S1(config)#vlan 10

```
S1(config-vlan10)#switchport interface ethernet 1/1
Set the port Ethernet1/1 access vlan 10 successfully
S1(config-vlan10)#exit
S1(config)#vlan 20
S1(config-vlan20)#switchport interface ethernet 1/2
Set the port Ethernet1/2 access vlan 20 successfully
S1(config-vlan20)#exit
S1(config)#interface ethernet 1/24
S1(config-if-ethernet1/24)#switchport mode trunk
Set the port Ethernet1/24 mode Trunk successfully
S1(config-if-ethernet1/24)#switchport trunk allowed vlan all
S1(config-if-ethernet1/24)#exit
S1(config)#
```

步骤 3：配置路由器，代码如下。

```
Router>enable
Router#config
Router_config#
Router_config#interface f0/0.1                    //配置子接口
Router_config_f0/0.1#encapsulation dot1Q 10       // 封装 802.1q
Router_config_f0/0.1#ip address 192.168.1.1 255.255.255.0    // 配置 VLAN10 的网关地址
Router_config_f0/0.1#exit
Router_config#interface f0/0.2
Router_config_f0/0.2#encapsulation dot1Q 20
Router_config_f0/0.2#ip address 192.168.2.1 255.255.255.0
Router_config_f0/0.2#exit
Router_config#
```

单臂路由配置完成，用 PC1 ping PC2，可以看到通信正常，代码如下。

```
C:\Documents and Settings\Administrator>ping 192.168.2.10

Pinging 192.168.2.10 with 32 bytes of data:

Reply from 192.168.2.10: bytes=32 time=1ms TTL=63
Reply from 192.168.2.10: bytes=32 time<1ms TTL=63
Reply from 192.168.2.10: bytes=32 time<1ms TTL=63
Reply from 192.168.2.10: bytes=32 time<1ms TTL=63

Ping statistics for 192.168.2.10:
    Packets: Sent = 4, Received = 4, Lost = 0 (0% loss),
```

Approximate round trip times in milli-seconds:
　　Minimum = 0ms, Maximum = 1ms, Average = 0ms

实训 2　三层交换机实现 VLAN 间互访

实训目标

李工：单臂路由方式受带宽限制，在大型网络中对网络性能影响较大。这时通常采用三层交换机。三层交换机在二层交换机基础上增加了网络层的功能，交换机以太网口较多，不受带宽限制，可用于较大型网络的 VLAN 间通信。

通过学习本实训，能够掌握通过三层交换机实现 VLAN 间互访，具备配置三层交换机 VLAN 的能力。

实训拓扑

实训拓扑图如图 3-2 所示。

图　3-2

实训任务

步骤 1：规划交换机端口、VLAN 及 IP 地址等，见表 3-2。

表　3-2

VLAN	端　口	VLAN 接口 IP	连接 PC	IP 地址	网　关
10	E1/1	192.168.1.1	PC1	192.168.1.10	192.168.1.1
20	E1/2	192.168.2.1	PC2	192.168.2.10	192.168.2.1

步骤 2：配置三层交换机。

划分 VLAN10 及 VLAN20，分别加入端口 E1/1 和 E1/2，配置各 VLAN 的接口地址，代码如下。

```
S1(config)#vlan 10
S1(Config-Vlan10)#switchport interface ethernet 1/1
Set the port Ethernet1/1 access vlan 10 successfully
S1(Config-Vlan10)#exit
S1(config)#vlan 20
```

```
S1(Config-Vlan20)#switchport interface ethernet 1/2
Set the port Ethernet1/2 access vlan 20 successfully
S1(Config-Vlan20)#exit
S1(config)#interface vlan 10
S1(Config-if-Vlan10)#ip address 192.168.1.1 255.255.255.0
S1(Config-if-Vlan10)#exit
S1(config)#interface vlan 20
S1(Config-if-Vlan20)#ip address 192.168.2.1 255.255.255.0
S1(Config-if-Vlan20)#exit
S1(config)#
```

查看交换机路由表,代码如下。

```
S1#show ip route
Codes: K - kernel, C - connected, S - static, R - RIP, B - BGP
       O - OSPF, IA - OSPF inter area
       N1 - OSPF NSSA external type 1, N2 - OSPF NSSA external type 2
       E1 - OSPF external type 1, E2 - OSPF external type 2
       i - IS-IS, L1 - IS-IS level-1, L2 - IS-IS level-2, ia - IS-IS inter area
       * - candidate default

C        127.0.0.0/8 is directly connected, Loopback
C        192.168.1.0/24 is directly connected, Vlan10
C        192.168.2.0/24 is directly connected, Vlan20
Total routes are : 3 item(s)
S1#
```

可以看到路由表中已包括直连路由 VLAN10 和 VLAN20 网段。

PC1 ping PC2,代码如下。

```
C:\Documents and Settings\Administrator>ping 192.168.2.10

Pinging 192.168.2.10 with 32 bytes of data:

Reply from 192.168.2.10: bytes=32 time=2ms TTL=63
Reply from 192.168.2.10: bytes=32 time<1ms TTL=63
Reply from 192.168.2.10: bytes=32 time<1ms TTL=63
Reply from 192.168.2.10: bytes=32 time<1ms TTL=63

Ping statistics for 192.168.2.10:
    Packets: Sent = 4, Received = 4, Lost = 0 (0% loss),
Approximate round trip times in milli-seconds:
    Minimum = 0ms, Maximum = 0ms, Average = 0ms
```

可以看到 PC1 与 PC2 通信正常。

实训 3　动态路由协议 RIP

RIP 路由协议

实训目标

李工：当两台以上三层交换机级联时，为了使各交换机上连接的不同网段之间能够互相通信，就需要配置动态路由协议。常见的动态路由协议有 RIP、OSPF 等。

RIP（Routing Information Protocol，路由信息协议）基于距离矢量算法，它使用"跳数"，即 Metric 来衡量到达目标地址的路由距离。这种协议只关心自己周围的世界，只与自己相邻的网络设备交换信息，范围限制在 15 跳之内，再远，它就不关心了。

通过学习本实训，能够掌握 RIP 协议，具备配置 RIP 路由功能的能力。

实训拓扑

实训拓扑图如图 3-3 所示。

图　3-3

实训任务

步骤 1：环境规划。

1）在交换机 S1 和交换机 S2 上分别划分基于端口的 VLAN，见表 3-3。

表　3-3

交 换 机	VLAN	端 口 成 员
交换机 S1	10	E1/1
	20	E1/2
	100	E1/24
交换机 S2	30	E1/1
	40	E1/2
	200	E1/24

2）交换机 S1 和 S2 通过的 24 口级联。

3）配置交换机 S1 和 S2 各 VLAN 接口的 IP 地址，见表 3-4。

表　3-4

VLAN10	VLAN20	VLAN30	VLAN40	VLAN100	VLAN200
192.168.10.1	192.168.20.1	192.168.30.1	192.168.40.1	192.168.100.1	192.168.100.2

4）PC1～PC4 的网络设置，见表 3-5。

表 3-5

设 备	IP 地址	网 关	掩 码
PC1	192.168.10.10	192.168.10.1	255.255.255.0
PC2	192.168.20.10	192.168.20.1	255.255.255.0
PC3	192.168.30.10	192.168.30.1	255.255.255.0
PC4	192.168.40.10	192.168.40.1	255.255.255.0

步骤 2：配置交换机 S1，代码如下。

```
S1(config)#vlan 10
S1(Config-Vlan10)#switchport interface ethernet 1/1
Set the port Ethernet1/1 access vlan 10 successfully
S1(Config-Vlan10)#exit
S1(config)#vlan 20
S1(Config-Vlan20)#switchport interface ethernet 1/2
Set the port Ethernet1/2 access vlan 20 successfully
S1(Config-Vlan20)#exit
S1(config)#vlan 100
S1(Config-Vlan100)#switchport interface ethernet 1/24
Set the port Ethernet1/24 access vlan 100 successfully
S1(Config-Vlan20)#exit
S1(config)#interface vlan 10
S1(Config-if-Vlan10)#ip address 192.168.10.1 255.255.255.0
S1(Config-if-Vlan10)#exit
S1(config)#interface vlan 20
S1(Config-if-Vlan20)#ip address 192.168.20.1 255.255.255.0
S1(Config-if-Vlan20)#exit
S1(config)#interface vlan 100
S1(Config-if-Vlan20)#ip address 192.168.100.1 255.255.255.0
S1(Config-if-Vlan20)#exit
S1(config)#
```

步骤 3：配置交换机 S2，代码如下。

```
S2(config)#vlan 30
S2(config-vlan30)#switchport interface ethernet 1/0/1
Set the port Ethernet1/0/1 access vlan 30 successfully
S2(config-vlan30)#exit
S2(config)#vlan 40
S2(config-vlan40)#switchport interface ethernet 1/0/2
Set the port Ethernet1/0/2 access vlan 40 successfully
S2(config-vlan40)#exit
S2(config)#vlan 200
```

```
S2(config-vlan200)#switchport interface ethernet 1/0/24
Set the port Ethernet1/0/24 access vlan 200 successfully
S2(config-vlan200)#exit
S2(config)#interface vlan 30
S2(config-if-vlan30)#ip address 192.168.30.1 255.255.255.0
S2(config-if-vlan30)#exit
S2(config)#interface vlan 40
S2(config-if-vlan40)#ip address 192.168.40.1 255.255.255.0
S2(config-if-vlan40)#exit
S2(config)#interface vlan 200
S2(config-if-vlan200)#ip address 192.168.100.2 255.255.255.0
S2(config-if-vlan200)#exit
S2(config)#
```

步骤 4：验证互通性，见表 3-6。

表 3-6

源 PC	端　口	目的 PC	端　口	结　果
PC1	E1/1	PC2	S1：E1/2	通
PC1	E1/1	VLAN100	S1：E1/24	通
PC1	E1/1	PC3	S2：E1/0/2	不通
PC1	E1/1	VLAN200	S2：E1/0/24	不通

查看 S1 路由表：

```
S1#show ip route
Codes: K - kernel, C - connected, S - static, R - RIP, B - BGP
       O - OSPF, IA - OSPF inter area
       N1 - OSPF NSSA external type 1, N2 - OSPF NSSA external type 2
       E1 - OSPF external type 1, E2 - OSPF external type 2
       i - IS-IS, L1 - IS-IS level-1, L2 - IS-IS level-2, ia - IS-IS inter area
       * - candidate default

C       127.0.0.0/8 is directly connected, Loopback
C       192.168.10.0/24 is directly connected, Vlan10
C       192.168.20.0/24 is directly connected, Vlan20
C       192.168.100.0/24 is directly connected, Vlan100
Total routes are : 4 item(s)
S1#
```

可以看到 S1 路由表中只有直连路由信息，即 192.168.10.0、192.168.20.0、192.168.100.0 网段，并没有 S2 交换机上的路由信息，因此 PC1 无法 ping 通 S2 交换机上的网段。

步骤 5：分别在交换机 S1 和 S2 上添加各自直连网段的动态路由协议 RIP，代码如下。

S1(config)#router rip
S1(config-router)#network vlan 10
S1(config-router)#network vlan 20
S1(config-router)#network vlan 100
S1(config-router)#exit
S1(config)#

S2(config)#router rip
S2(config-router)#network vlan 30
S2(config-router)#network vlan 40
S2(config-router)#network vlan 200
S2(config-router)#exit
S2(config)#

查看 S1 路由表：

S1#show ip route
Codes: K - kernel, C - connected, S - static, R - RIP, B - BGP
       O - OSPF, IA - OSPF inter area
       N1 - OSPF NSSA external type 1, N2 - OSPF NSSA external type 2
       E1 - OSPF external type 1, E2 - OSPF external type 2
       i - IS-IS, L1 - IS-IS level-1, L2 - IS-IS level-2, ia - IS-IS inter area
       * - candidate default

C   127.0.0.0/8 is directly connected, Loopback
C   192.168.10.0/24 is directly connected, Vlan10
C   192.168.20.0/24 is directly connected, Vlan20
R   192.168.30.0/24 [120/2] via 192.168.100.2, Vlan100, 00:03:05
R   192.168.40.0/24 [120/2] via 192.168.100.2, Vlan100, 00:03:05
C   192.168.100.0/24 is directly connected, Vlan100
Total routes are : 6 item(s)
S1#

验证互通性，见表 3-7。

表 3-7

源 PC	端口	目的 PC	端口	结果
PC1	E1/1	PC2	S1：E1/2	通
PC1	E1/1	VLAN100	S1：E1/24	通
PC1	E1/1	PC3	S2：E1/0/2	通
PC1	E1/1	VLAN200	S2：E1/0/24	通

实训 4　动态路由协议 OSPF

OSPF 路由协议

实训目标

李工：相对于 RIP，OSPF（Open Shortest Path First，开放式最短路径优先）协议是一种典型的链路状态（Link-state）的路由协议，一般用于同一个路由域内。OSPF 将链路状态广播数据（Link State Advertisement，LSA）传送给在某一区域内的所有路由器，这一点与 RIP 不同。因此，OSPF 协议能够计算出最优路径，收敛速度快，常应用于大型网络。

通过学习本实训，能够掌握 OSPF 协议，具备配置 OSPF 路由功能的能力。

实训拓扑

实训拓扑图如图 3-4 所示。

图 3-4

实训任务

步骤 1：环境规划。

1）在交换机 S1 和交换机 S2 上分别划分基于端口的 VLAN，见表 3-8。

表 3-8

交换机	VLAN	端口成员
交换机 S1	10	E1/1
	20	E1/2
	100	E1/24
交换机 S2	30	E1/1
	40	E1/2
	200	E1/24

2）交换机 S1 和 S2 通过的 24 口级联。

3）配置交换机 S1 和 S2 各 VLAN 接口的 IP 地址，见表 3-9。

表 3-9

VLAN10	VLAN20	VLAN30	VLAN40	VLAN100	VLAN200
192.168.10.1	192.168.20.1	192.168.30.1	192.168.40.1	192.168.100.1	192.168.100.2

4）PC1～PC4 的网络设置，见表 3-10。

表　3-10

设　　备	IP 地址	网　　关	掩　　码
PC1	192.168.10.10	192.168.10.1	255.255.255.0
PC2	192.168.20.10	192.168.20.1	255.255.255.0
PC3	192.168.30.10	192.168.30.1	255.255.255.0
PC4	192.168.40.10	192.168.40.1	255.255.255.0

步骤 2：配置交换机 S1，代码如下。

```
S1(config)#vlan 10
S1(Config-Vlan10)#switchport interface ethernet 1/1
Set the port Ethernet1/1 access vlan 10 successfully
S1(Config-Vlan10)#exit
S1(config)#vlan 20
S1(Config-Vlan20)#switchport interface ethernet 1/2
Set the port Ethernet1/2 access vlan 20 successfully
S1(Config-Vlan20)#exit
S1(config)#vlan 100
S1(Config-Vlan100)#switchport interface ethernet 1/24
Set the port Ethernet1/24 access vlan 100 successfully
S1(Config-Vlan20)#exit
S1(config)#interface vlan 10
S1(Config-if-Vlan10)#ip address 192.168.10.1 255.255.255.0
S1(Config-if-Vlan10)#exit
S1(config)#interface vlan 20
S1(Config-if-Vlan20)#ip address 192.168.20.1 255.255.255.0
S1(Config-if-Vlan20)#exit
S1(config)#interface vlan 100
S1(Config-if-Vlan20)#ip address 192.168.100.1 255.255.255.0
S1(Config-if-Vlan20)#exit
S1(config)#
```

步骤 3：配置交换机 S2，代码如下。

```
S2(config)#vlan 30
S2(config-vlan30)#switchport interface ethernet 1/0/1
Set the port Ethernet1/0/1 access vlan 30 successfully
S2(config-vlan30)#exit
S2(config)#vlan 40
S2(config-vlan40)#switchport interface ethernet 1/0/2
Set the port Ethernet1/0/2 access vlan 40 successfully
S2(config-vlan40)#exit
```

```
S2(config)#vlan 200
S2(config-vlan200)#switchport interface ethernet 1/0/24
Set the port Ethernet1/0/24 access vlan 200 successfully
S2(config-vlan200)#exit
S2(config)#interface vlan 30
S2(config-if-vlan30)#ip address 192.168.30.1 255.255.255.0
S2(config-if-vlan30)#exit
S2(config)#interface vlan 40
S2(config-if-vlan40)#ip address 192.168.40.1 255.255.255.0
S2(config-if-vlan40)#exit
S2(config)#interface vlan 200
S2(config-if-vlan200)#ip address 192.168.100.2 255.255.255.0
S2(config-if-vlan200)#exit
S2(config)#
```

步骤 4：验证互通性，见表 3-11。

表 3-11

源 PC	端口	目的 PC	端口	结果
PC1	E1/1	PC2	S1：E1/2	通
PC1	E1/1	VLAN100	S1：E1/24	通
PC1	E1/1	PC3	S2：E1/0/2	不通
PC1	E1/1	VLAN200	S2：E1/0/24	不通

查看 S1 路由表：

```
S1#show ip route
Codes: K - kernel, C - connected, S - static, R - RIP, B - BGP
       O - OSPF, IA - OSPF inter area
       N1 - OSPF NSSA external type 1, N2 - OSPF NSSA external type 2
       E1 - OSPF external type 1, E2 - OSPF external type 2
       i - IS-IS, L1 - IS-IS level-1, L2 - IS-IS level-2, ia - IS-IS inter area
       * - candidate default

C       127.0.0.0/8 is directly connected, Loopback
C       192.168.10.0/24 is directly connected, Vlan10
C       192.168.20.0/24 is directly connected, Vlan20
C       192.168.100.0/24 is directly connected, Vlan100
Total routes are : 4 item(s)
S1#
```

可以看到 S1 路由表中只有直连路由信息，即 192.168.10.0、192.168.20.0、192.168.100.0 网段，并没有 S2 上的路由信息，因此 PC1 无法 ping 通 S2 交换机上的网段。

步骤 5：分别在 S1 和 S2 上启动 OSPF 协议，并将对应的直连网段配置到 OSPF 进程中，代码如下。

```
S1(config)#router ospf
S1(config-router)#network 192.168.10.0/24 area 0
S1(config-router)#network 192.168.20.0/24 area 0
S1(config-router)#network 192.168.100.0/24 area 0
S1(config-router)#exit
S1(config)#

S2(config)#router ospf
S2(config-router)#network 192.168.30.0/24 area 0
S2(config-router)#network 192.168.40.0/24 area 0
S2(config-router)#network 192.168.100.0/24 area 0
S2(config-router)#exit
S2(config)#
```

查看 S1 路由表：

```
S1#show ip route
Codes: K - kernel, C - connected, S - static, R - RIP, B - BGP
       O - OSPF, IA - OSPF inter area
       N1 - OSPF NSSA external type 1, N2 - OSPF NSSA external type 2
       E1 - OSPF external type 1, E2 - OSPF external type 2
       i - IS-IS, L1 - IS-IS level-1, L2 - IS-IS level-2, ia - IS-IS inter area
       * - candidate default

C        127.0.0.0/8 is directly connected, Loopback  tag:0
C        192.168.10.0/24 is directly connected, Vlan10  tag:0
C        192.168.20.0/24 is directly connected, Vlan20  tag:0
O        192.168.30.0/24 [110/11] via 192.168.100.1, Vlan100, 00:01:10  tag:0
O        192.168.40.0/24 [110/11] via 192.168.100.1, Vlan100, 00:01:10  tag:0
C        192.168.100.0/24 is directly connected, Vlan100  tag:0
Total routes are : 6 item(s)
S1#
```

验证互通性，见表 3-12。

表 3-12

源 PC	端口	目的 PC	端口	结果
PC1	E1/1	PC2	S1：E1/2	通
PC1	E1/1	VLAN100	S1：E1/24	通
PC1	E1/1	PC3	S2：E1/0/2	通
PC1	E1/1	VLAN200	S2：E1/0/24	通

第 4 章　交换机高级应用实践

实训 1　标准访问控制列表

实训目标

ACL 访问控制列表

李工：访问控制列表（Access Control Lists, ACL）是一种数据包过滤机制，通过允许或拒绝特定的数据包进出网络，使交换机可以对网络访问进行控制，有效保证网络的安全运行。用户可以基于报文中的特定信息制定一组规则（Rule），每条规则都描述了对匹配一定信息的数据包所采取的动作：允许（Permit）或拒绝（Deny）通过。用户可以把这些规则应用到特定交换机端口的入口或出口方向，这样特定端口上特定方向的数据流就必须依照指定的 ACL 规则进出交换机。通过 ACL，可以限制某个 IP 地址的 PC 或者某些网段的 PC 的上网活动。

通过学习本实训，能够掌握标准访问控制列表的概念，具备配置 ACL 功能的能力。

实训拓扑

实训拓扑图如图 4-1 所示。

图　4-1

实训任务

步骤 1：环境规划。

1）在交换机 S1 和交换机 S2 上分别划分基于端口的 VLAN，见表 4-1。

表 4-1

交换机	VLAN	端口成员
交换机 S1	10	E0/0/1
	20	E0/0/2
	100	E0/0/24
交换机 S2	30	E0/0/1
	200	E0/0/24

2）交换机 S1 和 S2 通过的 24 口级联。

3）配置交换机 S1 和 S2 各 VLAN 接口的 IP 地址，见表 4-2。

表 4-2

VLAN10	VLAN20	VLAN30	VLAN100	VLAN200
192.168.10.1	192.168.20.1	192.168.30.1	192.168.100.1	192.168.100.2

4）PC1～PC3 的网络设置，见表 4-3。

表 4-3

设备	IP 地址	网关	掩码
PC1	192.168.10.10	192.168.10.1	255.255.255.0
PC2	192.168.20.10	192.168.20.1	255.255.255.0
PC3	192.168.30.10	192.168.30.1	255.255.255.0

步骤 2：配置交换机 VLAN 信息及虚接口地址。

交换机 S1：

```
S1(config)#vlan 10
S1(config-vlan10)#switchport interface ethernet 1/0/1
Set the port Ethernet1/0/1 access vlan 10 successfully
S1(config-vlan10)#exit
S1(config)#vlan 20
S1(config-vlan20)#switchport interface ethernet 1/0/2
Set the port Ethernet1/0/2 access vlan 20 successfully
S1(config-vlan20)#exit
S1(config)#vlan 100
S1(config-vlan20)#switchport interface ethernet 1/0/24
Set the port Ethernet1/0/24 access vlan 100 successfully
S1(config-vlan20)#exit
S1(config)#interface vlan 10
S1(config-if-vlan10)#ip address 192.168.10.1 255.255.255.0
S1(config-if-vlan10)#exit
S1(config)#interface vlan 20
S1(config-if-vlan20)#ip address 192.168.20.1 255.255.255.0
```

S1(config-if-vlan20)#exit

S1(config)#interface vlan 100

S1(config-if-vlan20)#ip address 192.168.100.1 255.255.255.0

S1(config-if-vlan20)#exit

S1(config)#

交换机 S2：

S2(config)#vlan 30

S2(Config-Vlan30)#switchport interface ethernet 1/1

Set the port Ethernet1/1 access vlan 30 successfully

S2(Config-Vlan30)#exit

S2(config)#vlan 200

S2(Config-Vlan30)#switchport interface ethernet 1/24

Set the port Ethernet1/24 access vlan 200 successfully

S2(Config-Vlan30)#exit

S2(config)# interface vlan 30

S2(Config-if-Vlan30)#ip address 192.168.30.1 255.255.255.0

S2(Config-if-Vlan30)#exit

S2(config)# interface vlan 200

S2(Config-if-Vlan30)#ip address 192.168.100.2 255.255.255.0

S2(Config-if-Vlan30)#exit

S2(config)#

步骤 3：配置静态路由。

交换机 S1：

S1(config)#ip route 0.0.0.0 0.0.0.0 192.168.100.2

S1(config)#show ip route

Codes: K - kernel, C - connected, S - static, R - RIP, B - BGP

     O - OSPF, IA - OSPF inter area

     N1 - OSPF NSSA external type 1, N2 - OSPF NSSA external type 2

     E1 - OSPF external type 1, E2 - OSPF external type 2

     i - IS-IS, L1 - IS-IS level-1, L2 - IS-IS level-2, ia - IS-IS inter area

     * - candidate default

Gateway of last resort is 192.168.100.2 to network 0.0.0.0

S*   0.0.0.0/0 [1/0] via 192.168.100.2, Vlan100   tag:0

C   127.0.0.0/8 is directly connected, Loopback   tag:0

C   192.168.10.0/24 is directly connected, Vlan10   tag:0

```
C        192.168.20.0/24 is directly connected, Vlan20    tag:0
C        192.168.100.0/24 is directly connected, Vlan100  tag:0
Total routes are : 4 item(s)
S1(config)#
```

交换机 S2：

```
S2(config)#ip route 0.0.0.0 0.0.0.0 192.168.100.1
S2(config)#show ip route
Codes: K - kernel, C - connected, S - static, R - RIP, B - BGP
       O - OSPF, IA - OSPF inter area
       N1 - OSPF NSSA external type 1, N2 - OSPF NSSA external type 2
       E1 - OSPF external type 1, E2 - OSPF external type 2
       i - IS-IS, L1 - IS-IS level-1, L2 - IS-IS level-2, ia - IS-IS inter area
       * - candidate default


Gateway of last resort is 192.168.100.1 to network 0.0.0.0

S*       0.0.0.0/0 [1/0] via 192.168.100.1, Vlan200
C        127.0.0.0/8 is directly connected, Loopback
C        192.168.30.0/24 is directly connected, Vlan30
C        192.168.100.0/24 is directly connected, Vlan200
Total routes are : 4 item(s)
S2(config)#
```

小张：师傅，这里可以使用 RIP 或 OSPF 动态路由实现吗？

李工：当然可以。

步骤 4：测试 PC1 与 PC3 的连通性，代码如下。

```
C:\Documents and Settings\Administrator>ping 192.168.30.10

Pinging 192.168.30.10 with 32 bytes of data:

Reply from 192.168.30.10: bytes=32 time=12ms TTL=62
Reply from 192.168.30.10: bytes=32 time<1ms TTL=62
Reply from 192.168.30.10: bytes=32 time<1ms TTL=62
Reply from 192.168.30.10: bytes=32 time<1ms TTL=62


Ping statistics for 192.168.30.10:
    Packets: Sent = 4, Received = 4, Lost = 0 (0% loss),
Approximate round trip times in milli-seconds:
    Minimum = 0ms, Maximum = 0ms, Average = 0ms
```

我们看到，这时 PC1 可以正常与 PC3 通信，PC2 同样可以与 PC3 通信。

步骤 5：在交换机 S2 上配置标准 ACL，代码如下。

```
S2(config)#access-list 10 deny 192.168.10.10 0.0.0.255      // 拒绝来自 192.168.10.0 网段的数据包
S2(config)#access-list 10 permit any-source                  // 允许其他任意网段的数据包通过
S2(config)#interface ethernet 1/0/24
S2(config-if-ethernet1/0/24)#ip access-group 10 in           // 将 ACL 应用到端口上
S2(config-if-ethernet1/0/24)#exit
S2(config)#
```

李工：小张，这样配置后会出现什么情况呢？

小张：师傅，是不是 PC2 可以 ping 通 PC3，但 PC1 则不能 ping 通 PC3 了吧。

李工：你可以试一试。

PC1 ping PC3：

```
C:\Documents and Settings\Administrator>ping 192.168.30.10

Pinging 192.168.30.10 with 32 bytes of data:

Request timed out.
Request timed out.
Request timed out.
Request timed out.

Ping statistics for 192.168.30.10:
    Packets: Sent = 4, Received = 0, Lost = 4 (100% loss),
```

PC2 ping PC3：

```
C:\Documents and Settings\Administrator>ping 192.168.30.10

Pinging 192.168.30.10 with 32 bytes of data:

Reply from 192.168.30.10: bytes=32 time=2ms TTL=62
Reply from 192.168.30.10: bytes=32 time<1ms TTL=62
Reply from 192.168.30.10: bytes=32 time<1ms TTL=62
Reply from 192.168.30.10: bytes=32 time<1ms TTL=62

Ping statistics for 192.168.30.10:
    Packets: Sent = 4, Received = 4, Lost = 0 (0% loss),
Approximate round trip times in milli-seconds:
    Minimum = 0ms, Maximum = 1ms, Average = 0ms
```

小张：果然是这样，S2 拒绝了来自 PC1 的数据包。

实训 2　扩展访问控制列表

ACL 访问控制列表

实训目标

李工：标准 ACL 只能限制源 IP 地址，在目的端拒绝源 IP 地址的访问。而扩展 ACL 的限制权限就很广泛，包括源 IP、目的 IP、服务类型等。下面以限制 ping 包 ICMP 为例，介绍扩展 ACL 的应用。

通过学习本实训，能够掌握扩展访问控制列表的概念，具备配置扩展 ACL 功能的能力。

实训拓扑

实训拓扑图如图 4-2 所示。

图 4-2

实训任务

步骤 1：环境规划。

1）在交换机 S1 和交换机 S2 上分别划分基于端口的 VLAN，见表 4-4。

表 4-4

交换机	VLAN	端口成员
交换机 S1	10	E0/0/1
	20	E0/0/2
	100	E0/0/24
交换机 S2	30	E0/0/1
	200	E0/0/24

2）交换机 S1 和 S2 通过的 24 口级联。

3）配置交换机 S1 和 S2 各 VLAN 接口的 IP 地址，见表 4-5。

表 4-5

VLAN10	VLAN20	VLAN30	VLAN100	VLAN200
192.168.10.1	192.168.20.1	192.168.30.1	192.168.100.1	192.168.100.2

4）PC1～PC3 的网络设置，见表 4-6。

表 4-6

设备	IP 地址	网关	掩码
PC1	192.168.10.10	192.168.10.1	255.255.255.0
PC2	192.168.20.10	192.168.20.1	255.255.255.0
PC3	192.168.30.10	192.168.30.1	255.255.255.0

步骤2：配置交换机 VLAN 信息及虚接口地址。

交换机 S1：

S1(config)#vlan 10

S1(config-vlan10)#switchport interface ethernet 1/0/1

Set the port Ethernet1/0/1 access vlan 10 successfully

S1(config-vlan10)#exit

S1(config)#vlan 20

S1(config-vlan20)#switchport interface ethernet 1/0/2

Set the port Ethernet1/0/2 access vlan 20 successfully

S1(config-vlan20)#exit

S1(config)#vlan 100

S1(config-vlan20)#switchport interface ethernet 1/0/24

Set the port Ethernet1/0/24 access vlan 100 successfully

S1(config-vlan20)#exit

S1(config)#interface vlan 10

S1(config-if-vlan10)#ip address 192.168.10.1 255.255.255.0

S1(config-if-vlan10)#exit

S1(config)#interface vlan 20

S1(config-if-vlan20)#ip address 192.168.20.1 255.255.255.0

S1(config-if-vlan20)#exit

S1(config)#interface vlan 100

S1(config-if-vlan20)#ip address 192.168.100.1 255.255.255.0

S1(config-if-vlan20)#exit

S1(config)#

交换机 S2：

S2(config)#vlan 30

S2(Config-Vlan30)#switchport interface ethernet 1/1

Set the port Ethernet1/1 access vlan 30 successfully

S2(Config-Vlan30)#exit

S2(config)#vlan 200

S2(Config-Vlan30)#switchport interface ethernet 1/24

Set the port Ethernet1/24 access vlan 200 successfully

S2(Config-Vlan30)#exit

S2(config)# interface vlan 30

S2(Config-if-Vlan30)#ip address 192.168.30.1 255.255.255.0

S2(Config-if-Vlan30)#exit

S2(config)# interface vlan 200

S2(Config-if-Vlan30)#ip address 192.168.100.2 255.255.255.0

S2(Config-if-Vlan30)#exit
S2(config)#

步骤 3：配置静态路由。

交换机 S1：

S1(config)#ip route 0.0.0.0 0.0.0.0 192.168.100.2
S1(config)#show ip route
Codes: K - kernel, C - connected, S - static, R - RIP, B - BGP
   O - OSPF, IA - OSPF inter area
   N1 - OSPF NSSA external type 1, N2 - OSPF NSSA external type 2
   E1 - OSPF external type 1, E2 - OSPF external type 2
   i - IS-IS, L1 - IS-IS level-1, L2 - IS-IS level-2, ia - IS-IS inter area
   * - candidate default

Gateway of last resort is 192.168.100.2 to network 0.0.0.0

S* 0.0.0.0/0 [1/0] via 192.168.100.2, Vlan100 tag:0
C 127.0.0.0/8 is directly connected, Loopback tag:0
C 192.168.10.0/24 is directly connected, Vlan10 tag:0
C 192.168.20.0/24 is directly connected, Vlan20 tag:0
C 192.168.100.0/24 is directly connected, Vlan100 tag:0
Total routes are : 4 item(s)
S1(config)#

交换机 S2：

S2(config)#ip route 0.0.0.0 0.0.0.0 192.168.100.1
S2(config)#show ip route
Codes: K - kernel, C - connected, S - static, R - RIP, B - BGP
   O - OSPF, IA - OSPF inter area
   N1 - OSPF NSSA external type 1, N2 - OSPF NSSA external type 2
   E1 - OSPF external type 1, E2 - OSPF external type 2
   i - IS-IS, L1 - IS-IS level-1, L2 - IS-IS level-2, ia - IS-IS inter area
   * - candidate default

Gateway of last resort is 192.168.100.1 to network 0.0.0.0

S* 0.0.0.0/0 [1/0] via 192.168.100.1, Vlan200
C 127.0.0.0/8 is directly connected, Loopback
C 192.168.30.0/24 is directly connected, Vlan30
C 192.168.100.0/24 is directly connected, Vlan200

```
Total routes are : 4 item(s)
S2(config)#
```

步骤 4：测试 PC1、PC2 到 PC3 的连通性，代码如下。

```
C:\Documents and Settings\Administrator>ping 192.168.30.10

Pinging 192.168.30.10 with 32 bytes of data:

Reply from 192.168.30.10: bytes=32 time=12ms TTL=62
Reply from 192.168.30.10: bytes=32 time<1ms TTL=62
Reply from 192.168.30.10: bytes=32 time<1ms TTL=62
Reply from 192.168.30.10: bytes=32 time<1ms TTL=62

Ping statistics for 192.168.30.10:
    Packets: Sent = 4, Received = 4, Lost = 0 (0% loss),
Approximate round trip times in milli-seconds:
    Minimum = 0ms, Maximum = 0ms, Average = 0ms
```

可以看到，这时 PC1 与 PC2 都可以正常 ping 通 PC3。

步骤 5：在交换机 S1 上配置扩展 ACL，代码如下。

```
S1(config)#access-list 101 deny icmp 192.168.20.10 0.0.0.255 192.168.30.10 0.0.0.255
            // 拒绝 192.168.20.0 网段的 ICMP 包访问 192.168.30.0 网段
S1(config)#access-list 101 permit icmp any-source any-destination
            // 允许来自其他网段的 ICMP 包访问任意网段
S1(config)#interface ethernet 1/0/1-2
S1(config-if-port-range)#ip access-group 101 in         // 将 ACL 应用到端口上
S1(config-if-port-range)#exit
S1(config)#
```

小张：为什么这里要配置在 S1 上呢？

李工：扩展 ACL 应尽量靠近源地址，所以这里把扩展 ACL 配置在 S1 上。而标准 ACL 由于只能限制源 IP 地址，为了避免错误的限制，应尽量靠近目的地址，故第 4 章实训 1 中配置在了 S2 上。

现在用 PC1 ping PC3，应该是通的，代码如下。

```
C:\Documents and Settings\Administrator>ping 192.168.30.10

Pinging 192.168.30.10 with 32 bytes of data:

Reply from 192.168.30.10: bytes=32 time=1ms TTL=62
Reply from 192.168.30.10: bytes=32 time<1ms TTL=62
Reply from 192.168.30.10: bytes=32 time<1ms TTL=62
Reply from 192.168.30.10: bytes=32 time<1ms TTL=62
```

Ping statistics for 192.168.30.10:

    Packets: Sent = 4, Received = 4, Lost = 0 (0% loss),

Approximate round trip times in milli-seconds:

    Minimum = 0ms, Maximum = 1ms, Average = 0ms

而用 PC2 ping PC3，则不能 ping 通，代码如下。

C:\Documents and Settings\Administrator>ping 192.168.30.10

Pinging 192.168.30.10 with 32 bytes of data:

Request timed out.
Request timed out.
Request timed out.
Request timed out.

Ping statistics for 192.168.30.10:

    Packets: Sent = 4, Received = 0, Lost = 4 (100% loss),

实训 3　三层交换机 DHCP 服务

实 训 目 标

李工：网络中一般都采用 DHCP 作为地址分配的方法，以减轻网络管理员和用户的配置负担。这里可以将支持 DHCP 的交换机配置成 DHCP 服务器，自动为连接它的终端设备分配 IP 地址。

通过学习本实训，能够掌握 DHCP 服务，具备配置三层交换机 DHCP 服务的能力。

实 训 拓 扑

实训拓扑图如图 4-3 所示。

图　4-3

实训任务

步骤1：环境规划。

划分两个 VLAN，见表 4-7。

表 4-7

VLAN	IP 地址	端口成员
VLAN10	192.168.10.1/24	1、2
VLAN20	192.168.20.1/24	3

配置两个地址池，见表 4-8。

表 4-8

POOLA （192.168.10.0/24）		POOLB （192.168.20.0/24）	
设备 PC1	IP 自动获取	设备 PC3	IP 自动获取
默认网关	192.168.10.1	默认网关	192.168.20.1
DNS 服务器	192.168.1.1	DNS 服务器	192.168.1.1
Lease	8 小时	Lease	1 小时

其中在 VLAN10 中，需将 MAC 地址为 00-0B-4C-9E-2A-1C 的 PC2 分配固定 IP 地址 192.168.10.88。

步骤2：划分 VLAN，加入端口，配置 VLAN 虚接口地址，代码如下。

```
switch(config)#vlan 10
switch(config-vlan10)#switchport interface ethernet 1/0/1-2
Set the port Ethernet1/0/1 access vlan 10 successfully
Set the port Ethernet1/0/2 access vlan 10 successfully
switch(config-vlan10)#exit
switch(config)#vlan 20
switch(config-vlan20)#switchport interface ethernet 1/0/3
Set the port Ethernet1/0/3 access vlan 20 successfully
switch(config-vlan20)#exit
switch(config)# interface vlan 10
switch(config-if-vlan10)#ip address 192.168.10.1 255.255.255.0
switch(config-if-vlan10)#exit
switch(config)#interface vlan 20
switch(config-if-vlan20)#ip address 192.168.20.1 255.255.255.0
switch(config-if-vlan20)#exit
switch(config)#
```

步骤3：配置 DHCP，代码如下。

```
switch(config)#service dhcp
switch(config)#ip dhcp pool POOLA
switch(dhcp-poola-config)#network-address 192.168.10.0 24
switch(dhcp-poola-config)#lease 0 8
switch(dhcp-poola-config)#default-router 192.168.10.1
switch(dhcp-poola-config)#dns-server 192.168.1.1
switch(dhcp-poola-config)#exit
switch(config)#ip dhcp pool POOLB
switch(dhcp-poolb-config)#network-address 192.168.20.0 24
switch(dhcp-poolb-config)#lease 0 1
switch(dhcp-poolb-config)#default-router 192.168.20.1
switch(dhcp-poolb-config)#dns-server 192.168.1.1
switch(dhcp-poolb-config)#exit
switch(config)#
```

步骤4：为特殊的 PC2 配置地址池，代码如下。

```
switch(config)#ip dhcp excluded-address 192.168.10.80 192.168.10.90
switch(config)#ip dhcp pool POOLC
switch(dhcp-poolc-config)#host 192.168.10.88
switch(dhcp-poolc-config)#hardware-address 00-0B-4C-9E-2A-1C
switch(dhcp-poolc-config)#default-router 192.168.10.1
switch(dhcp-poolc-config)#dns-server 192.168.1.1
switch(dhcp-poolc-config)#exit
switch(config)#
```

步骤5：验证，见表 4-9。

表 4-9

设 备	端 口	结 果
PC1	E0/0/1	192.168.10.2/24
PC2	E0/0/2	192.168.10.88/24
PC3	E0/0/3	192.168.20.2/24

实训 4　三层交换机 DHCP 中继

实训目标

李工：当 DHCP 客户机和 DHCP 服务器不在同一个网段时，由 DHCP 中继传递 DHCP 报文。增加 DHCP 中继功能的好处是不必为每个网段都设置 DHCP 服务器，同一个 DHCP 服务器可以为很多个子网的客户机提供网络配置参数，既节约了成本，又方便了管理。这就是 DHCP 中继的功能。

通过学习本实训，能够掌握 DHCP 中继服务，具备配置三层交换机 DHCP 中继的能力。

实训拓扑

实训拓扑图如图 4-4 所示。

图 4-4

实训任务

步骤 1：环境规划。

交换机 S1 为 DHCP 服务器，划分两个 VLAN，见表 4-10。

表 4-10

VLAN	IP 地址	端口成员
VLAN10	192.168.10.1/24	1
VLAN100	192.168.100.1/24	24

为三个 VLAN 分别配置三个地址池。

交换机 S2 开启 DHCP 中继，划分三个 VLAN，见表 4-11。

表 4-11

VLAN	IP 地址	端口成员
VLAN20	192.168.20.1/24	1
VLAN30	192.168.30.1/24	2
VLAN200	192.168.100.2/24	24

其中，两台交换机通过 24 口相连，并配置为 192.168.100.0 网段。

步骤 2：分别在两台交换机上划分 VLAN，加入端口，配置虚接口地址。

交换机 S1：

```
S1(config)#vlan 10
S1(Config-Vlan10)#switchport interface ethernet 1/1
Set the port Ethernet1/1 access vlan 10 successfully
S1(Config-Vlan10)#exit
S1(config)#vlan 100
S1(Config-Vlan100)#switchport interface ethernet 1/24
Set the port Ethernet1/24 access vlan 100 successfully
S1(Config-Vlan100)#exit
S1(config)#int vlan 10
S1(Config-if-Vlan10)#ip address 192.168.10.1 255.255.255.0
S1(Config-if-Vlan10)#exit
```

```
S1(config)#int vlan 100
S1(Config-if-Vlan100)#ip address 192.168.100.1 255.255.255.0
S1(Config-if-Vlan100)#exit
```

交换机 S2：

```
S2(config)#vlan 20
S2(config-vlan20)#switchport interface ethernet 1/0/1
Set the port Ethernet1/0/1 access vlan 20 successfully
S2(config-vlan20)#exit
S2(config)#vlan 30
S2(config-vlan30)#switchport interface ethernet 1/0/2
Set the port Ethernet1/0/2 access vlan 30 successfully
S2(config-vlan30)#exit
S2(config)#vlan 200
S2(config-vlan200)#switchport interface ethernet 1/0/24
Set the port Ethernet1/0/24 access vlan 200 successfully
S2(config-vlan200)#exit
S2(config)#int vlan 20
S2(Config-if-Vlan20)#ip address 192.168.20.1 255.255.255.0
S2(Config-if-Vlan20)#exit
S2(config)#int vlan 30
S2(Config-if-Vlan30)#ip address 192.168.30.1 255.255.255.0
S2(Config-if-Vlan30)#exit
S2(config)#int vlan 200
S2(Config-if-Vlan200)#ip address 192.168.100.2 255.255.255.0
S2(Config-if-Vlan200)#exit
```

步骤 3：配置静态路由。

交换机 S1：

```
S1(config)#ip route 0.0.0.0 0.0.0.0 192.168.100.2
```

交换机 S2：

```
S2(config)#ip route 0.0.0.0 0.0.0.0 192.168.100.1
```

步骤 4：配置交换机 S1 的 DHCP 地址池，代码如下。

```
S1(config)#service dhcp
S1(config)#ip dhcp pool poola
S1(dhcp-poola-config)#network-address 192.168.10.0 24
S1(dhcp-poola-config)#default-router 192.168.10.1
S1(dhcp-poola-config)#dns-server 192.168.1.1
S1(dhcp-poola-config)#exit
S1(config)#ip dhcp pool poolb
S1(dhcp-poolb-config)#network-address 192.168.20.0 24
```

```
S1(dhcp-poolb-config)#default-router 192.168.20.1
S1(dhcp-poolb-config)#dns-server 192.168.1.1
S1(dhcp-poolb-config)#exit
S1(config)#ip dhcp pool poolc
S1(dhcp-poolc-config)#network-address 192.168.30.0 24
S1(dhcp-poolc-config)#default-router 192.168.30.1
S1(dhcp-poolc-config)#dns-server 192.168.1.1
S1(dhcp-poolc-config)#exit
```

步骤 5：配置交换机 S2 的 DHCP 中继，代码如下。

```
S2(config)#service dhcp
S2(config)#ip forward-protocol udp bootps
S2(config)#int vlan 20
S2(Config-if-vlan20)#ip helper-address 192.168.100.1
S2(Config-if-vlan20)#exit
S2(Config)#int vlan 30
S2(config-if-vlan30)#ip helper-address 192.168.100.1
S2(config-if-vlan30)#exit
S2(config)#
```

步骤 6：验证，见表 4-12。

表 4-12

设 备	端 口	结 果
PC1	S1 E1/1	192.168.10.2/24
PC2	S2 E1/0/1	192.168.20.2/24
PC3	S2 E1/0/2	192.168.30.2/24

实训 5　虚拟路由器冗余协议

实训目标

李工：虚拟路由器冗余协议（Virtual Router Redundancy Protocol，VRRP）是一种容错协议，运行于局域网的多台路由器上，它将这几台路由器组织成一台"虚拟"路由器，或称为一个备份组（Standby Group）。在 VRRP 备份组内，总有一台路由器或以太网交换机是活动路由器（Master），它完成"虚拟"路由器的工作；该备份组中其他的路由器或以太网交换机作为备份路由器（Backup，可以不只一台），随时监控 Master 的活动。当原有的 Master 出现故障时，各 Backup 将自动选举出一个新的 Master 来接替其工作，继续为网段内各主机提供路由服务。由于这个选举和接替阶段短暂而平滑，因此，网段内各主机仍然可以正常地使用虚拟路由器，实现不间断地与外界保持通信。

通过学习本实训，能够掌握 VRRP 协议，具备配置 VRRP 功能的能力。

实训拓扑

实训拓扑图如图 4-5 所示。

图 4-5

实训任务

步骤 1：环境规划。

在 S1 和 S2 上划分 VLAN，见表 4-13。

表 4-13

交换机	VLAN	端口成员	IP 地址
S1	100	1	10.0.0.1/24
	200	24	20.0.0.1/24
S2	100	1	10.0.0.2/24
	200	24	20.0.0.2/24

PC1 和 PC2 的网络设置，见表 4-14。

表 4-14

设备	IP 地址	网关（VRRP 虚拟网关）	掩码
PC1	10.0.0.3	10.0.0.10	255.255.255.0
PC2	20.0.0.3	20.0.0.10	255.255.255.0

步骤 2：在交换机 S1 和 S2 上分别划分 VLAN，加入端口，配置虚接口地址。

交换机 S1：

S1(config)#vlan 100

S1(config-vlan100)#switchport interface ethernet 1/0/1

Set the port Ethernet1/0/1 access vlan 100 successfully

S1(config-vlan100)#exit

S1(config)#vlan 200

S1(config-vlan200)#switchport interface ethernet 1/0/24

Set the port Ethernet1/0/24 access vlan 200 successfully

S1(config-vlan200)#exit

S1(config)#int vlan 100

S1(config-if-vlan100)#ip address 10.0.0.1 255.255.255.0

```
S1(config-if-vlan100)#exit
S1(config)#int vlan 200
S1(config-if-vlan200)#ip address 20.0.0.1 255.255.255.0
S1(config-if-vlan200)#exit
S1(config)#
```

交换机 S2：

```
S2(config)#vlan 100
S2(Config-Vlan100)#switchport interface ethernet 1/1
Set the port Ethernet1/1 access vlan 100 successfully
S2(Config-Vlan100)#exit
S2(config)#vlan 200
S2(Config-Vlan200)#switchport interface ethernet 1/24
Set the port Ethernet1/24 access vlan 200 successfully
S2(Config-Vlan200)#exit
S2(config)#int vlan 100
S2(Config-if-Vlan100)#ip address 10.0.0.2 255.255.255.0
S2(Config-if-Vlan100)#exit
S2(config)#int vlan 200
S2(Config-if-Vlan200)#ip address 20.0.0.2 255.255.255.0
S2(Config-if-Vlan200)#exit
S2(config)#
```

步骤 3：配置 VRRP。

交换机 S1：

```
S1(config)#router vrrp 1
S1(config-router)#virtual-ip 10.0.0.10
S1(config-router)#priority 150
S1(config-router)#interface vlan 100
S1(config-router)#circuit-failover vlan 200 110
S1(config-router)#enable
S1(config-router)#exit
S1(config)#router vrrp 2
S1(config-router)#virtual-ip 20.0.0.10
S1(config-router)#priority 50
S1(config-router)#interface vlan 200
S1(config-router)#enable
S1(config-router)#exit
S1(config)#
```

验证配置：

S1#show vrrp
VrId 1
 State is Master
 Virtual IP is 10.0.0.10 (Not IP owner)
 Interface is Vlan100
 Configured priority is 150, Current priority is 150
 Advertisement interval is 1 sec
 Preempt mode is TRUE
 Circuit failover interface Vlan200, Priority Delta 110, Status UP
VrId 2
 State is Backup
 Virtual IP is 20.0.0.10 (Not IP owner)
 Interface is Vlan200
 Priority is 50
 Advertisement interval is 1 sec
 Preempt mode is TRUE

S1#

交换机 S2：

S2(config)#router vrrp 1
S2(config-router)#virtual-ip 10.0.0.10
S2(config-router)#priority 50
S2(config-router)#interface vlan 100
S2(config-router)#enable
S2(config-router)#exit
S2(config)#router vrrp 2
S2(config-router)#virtual-ip 20.0.0.10
S2(config-router)#priority 150
S2(config-router)#interface vlan 200
S2(config-router)#circuit-failover vlan 100 110
S2(config-router)#enable
S2(config-router)#exit
S2(config)#

验证配置：

S2#show vrrp
VrId 1
 State is Backup

Virtual IP is 10.0.0.10 (Not IP owner)

Interface is Vlan100

Priority is 50

Advertisement interval is 1 sec

Preempt mode is TRUE

VrId 2

State is Master

Virtual IP is 20.0.0.10 (Not IP owner)

Interface is Vlan200

Configured priority is 150, Current priority is 150

Advertisement interval is 1 sec

Preempt mode is TRUE

Circuit failover interface Vlan100, Priority Delta 110, Status UP

S2#

步骤 4：验证。

PC1 持续 ping PC2，分别断开 S1 到 S3 及 S1 到 S4 的链路，如图 4-6 所示。

图 4-6

可以看到 VRRP 状态切换时有少量丢包，工作正常。

第2部分

路由器实训

第 5 章　路由器实践基础

实训 1　路由器的基本管理方法

路由器设备简介

实 训 目 标

小刘中职毕业后经过了集成公司的历练，终于盼来了给运营商做网络的机会，可师傅告诉他，运营商里最低级别都是企业级的路由器，让他抓紧时间好好学习一下路由器的知识。由于下周就要开始项目的实施，小刘被告知务必要在 3 天的时间里搞清楚路由器的带内管理方法和带外管理方法，并尽可能多地了解路由器的端口特性和配置模式等常识。

通过学习本实训，熟悉路由器各接口和作用，具备管理路由器的能力。

实 训 设 备

1）DCR-2655/2659/2855 路由器 1 台。
2）PC 1 台。
3）Console 线缆、交叉双绞线各 1 条。

实 训 拓 扑

实训拓扑图如图 5-1 所示。

图　5-1

实 训 任 务

步骤 1：通过 Console 口对设备进行初始配置（带外管理方法）。

1）将配置线的一端与路由器的 Console 口相连，另一端与 PC 的串口相连，如图 5-1 所示。

2）在 PC 上运行终端仿真程序。执行"开始"→"程序"→"附件"→"通信"→"超级终端"命令，打开如图 5-2 所示的对话框，设置终端的硬件参数如下（包括串口号）。

每秒位数：9600。

数据位：8。

奇偶校验：无。

停止位：1。

数据流控制：无。

图 5-2

3）路由器加电，超级终端会显示路由器自检信息，自检结束后出现命令提示"Press RETURN to get started"，代码如下。

```
System Bootstrap, Version 0.4.2        // 路由器版本信息
Serial num:8IRTJ610B407000041, ID num:200213
Copyright 2011 by Digital China Networks(BeiJing) Limited
Digital China-DCR-2659 Series DCR-2659        // 路由器型号
The current time: 65024-118-60 2:227:120
Loading DCR26V1.3.3H.bin......        // 加载配置文件
Start Decompress DCR26V1.3.3H.bin
################################################################################
################################################################################
################################################################################
Decompress 5519234 byte,Please wait system up..
Digital China Networks Limited Internetwork Operating System Software
Digital China-DCR-2659 Series Software , Version 1.3.3H, RELEASE SOFTWARE
System start up OK
Router console 0 is now available
Press RETURN to get started
```

4）按〈Enter〉键进入用户配置模式。DCR 路由器出厂时没有设定密码，用户按〈Enter〉键直接进入普通用户模式。用户可以使用权限允许范围内的命令，需要帮助可以随时输入"？"，输入"enable"，按〈Enter〉键则进入特权模式。这时候用户拥有最大权限，可以任意配置，需要帮助可以随时输入"？"。

```
Router>
Router>enable                                         // 进入特权模式
Router#Jan  1 00:03:58 Unknown user enter privilege mode from console 0, level = 15
Router#?                                              // 查看可用命令
  cd              -- Change directory
  chinese         -- Help message in Chinese
  chmem           -- Change memory of system
  cham            -- Change memory
  clear           -- Clear something
  config          -- Enter configurative mode
  connect         -- Open a outgoing connection
  copy            -- Copy configuration or image data
  debug           -- Debugging functions
  delete          -- Delete a file
  dir             -- List files in flash memory
  disconnect      -- Disconnect an existing outgoing network connection
  download        -- Download with ZMODEM
  enable          -- Turn on privileged commands
  english         -- Help message in English
  exec-script     -- Execute a script on a port or line
  exit            -- Exit / quit
  format          -- Format file system
  help            -- Description of the interactive help system
  history         -- Look up history
  keepalive       -- Keepalive probe
  look            -- Display memory
 --More—                                              // 按〈Enter〉键可以查看更多命令
Router#ch?                                            // 使用？帮助
  chinese         -- Help message in Chinese
  chmem           -- Change memory of system
  cham            -- Change memory
Router#chinese                                        // 使用中文帮助
Router#?                                              // 再次查看可用命令
  cd              -- 改变当前目录
```

命令	说明
chinese	-- 中文帮助信息
chmem	-- 修改系统内存数据
cham	-- 修改内存数据
clear	-- 清除
config	-- 进入配置状态
connect	-- 打开一个向外的连接
copy	-- 复制配置方案或内存映像
debug	-- 分析功能
delete	-- 删除一个文件
dir	-- 显示闪存中的文件
disconnect	-- 断开活跃的网络连接
download	-- 通过 ZMODEM 协议下载文件
enable	-- 进入特权方式
english	-- 英文帮助信息
exec-script	-- 在指定端口运行指定的脚本
exit	-- 退回或退出
format	-- 格式化文件系统
help	-- 交互式帮助系统描述
history	-- 查看历史
keepalive	-- 保活探测
look	-- 显示内存数据

--More--

步骤2：给相应的接口配置 IP 地址，开启相应的服务，进行带内管理。

1）设置路由器以太网接口地址并验证连通性，代码如下。

```
Router>enable                      // 进入特权模式
Router#Jan  1 00:11:17 Unknown user enter privilege mode from console 0, level = 15
Router#config                      // 进入全局配置模式
Router_config#interface f0/0       // 进入接口模式
Router_config_f0/0#ip address 192.168.2.1 255.255.255.0   // 配置 IP 地址
Router_config_f0/0#no shutdown     // 开启端口
Router_config_f0/0#^Z              // 返回特权模式 （Ctrl+Z）
Router#Jan  1 00:12:15 Configured from console 0 by


Router#show interface f0/0                    // 验证
FastEthernet0/0 is up, line protocol is up    // 接口和协议都必须是 up
address is 00e0.0f26.0381
  MTU 1500 bytes, BW 100000 kbit, DLY 10 usec
  Interface address is 192.168.2.1/24                  // 接口地址为 192.168.2.1/24
```

```
Encapsulation ARPA
ARP type: ARPA, ARP timeout 00:03:00
60 second input rate 151 bits/sec, 0 packets/sec!
60 second output rate 6 bits/sec, 0 packets/sec!
Full-duplex, 100Mb/s, 100BaseTX, 36 ii, 1 oi        // 全双工、百兆
 36 packets input, 6394 bytes, 200 rx_freebuf
 Received 0 unicasts, 0 lowmark, 36 ri, 0 input errors
 0 overrun, 0 CRC, 0 framing, 0 busy, 0 long, 0 discard, 0 throttles
 1 packets output, 46 bytes, 210 tx_freebd, 0 output errors
 0 underrun, 0 collisions, 0 late collisions, 0 deferred, 0 reTx expired
 0 resets, 0 lost carrier, 0 no carrier 0 grace stop 0 bus error
 0 output buffer failures, 0 output buffers swapped out 0 tx errors
 0 packets fast-switch
```

2) 设置 PC 的 IP 地址并测试连通性，如图 5-3 所示。

图 5-3

3) 使用 ping 测试连通性，如图 5-4 所示。

图 5-4

4）在 PC 上 Telnet 登录路由器。

为了保证安全性，路由器中默认为所有的 Telnet 用户必须通过验证才可以进入路由器配置页面，在 DCR 系列路由器中，系统支持使用本地或其他验证数据库对 Telnet 用户进行验证，具体过程可参考如下：

设置本地数据库中的用户名，本例使用的用户名和密码均为 dcnu。

Router_config #*username dcnu password dcnu*　// 设置本地用户名和密码
Router_config#

创建一个新的登录验证方法，名为 login_fortelnet，此方法将使用本地数据库验证。

Router_config#*aaa authentication login login_fortelnet local*　// 创建 login_fortelnet 验证，采用 local
Router_config#

进入 Telnet 进程管理配置模式，配置登录用户使用 login_fortelnet 的验证方法进行验证。

Router_config#*line vty 0 4*
Router_config_line#*login authentication login_fortelnet*　// 在接口下应用
Router_config_line#

经过这样的配置，Telnet 登录路由器时的过程将如下所示：

C:\>*telnet 192.168.2.1*
User Access Verification
Username: dcnu
Password:
　　　　Welcome to Digital China Multi-Protocol DCR-2659 Series
Router>

注意事项和排错

1）在超级终端中的配置是对路由器的操作，这时的 PC 只是输入输出设备。
2）在 Telnet 和 Web 方式管理时，先测试连通性。

相关命令详解

相关命令详解，见表 5-1。

表 5-1

命　令	解　释
interface fastethernet 0/0	进入快速以太网口 F0/0
ip address 192.168.2.2 255.255.255.0	设置 IP 地址为 192.168.2.2，子网掩码为 255.255.255.0
no shutdown	激活端口
ip http server	启动 HTTP 服务
username dcnu password dcnu	用户名：dcnu，密码：dcnu
aaa authentication login login_forhttp local	创建 login_fortelnet 验证，采用 local

共同思考

1）带内管理和带外管理各有什么优点和缺点？
2）Telnet 和 Web 的端口号是什么？

实训 2　维护路由器的配置文件

实训目标

小刘通过三天时间的熟悉，终于弄懂了企业级路由器的带内管理和带外管理，可是他又发现了新的问题，就是如何对路由器的软件进行升级，以及如何对路由器的配置文件进行备份和还原，于是他决定利用一天的时间来搞懂这两个问题。

通过学习本实训，能够掌握路由器配置文件的管理，具备维护路由器的能力。

实训设备

1）DCR-2655/2659/2855 路由器 1 台。
2）PC 1 台。
3）TFTP、FTP 软件。
4）Console 线缆、交叉双绞线各 1 条。

实训拓扑

实训拓扑图如图 5-5 所示。

图　5-5

实训任务

步骤 1：TFTP 方式（采用 UDP，适合本地操作）。

1）设置 PC 网卡地址为 192.168.2.10，并安装 TFTP 服务器（方法参考第 1 章实训 3）。
2）设置 DCR-2655/2659/2855 的 F0/0 接口地址为 192.168.2.1，并测试连通性，代码如下。

```
Router>enable        //进入特权模式
Router#config        //进入全局模式
```

```
Router_config#hostname R1          // 修改主机名为 R1
R1_config_f0/0#ip address 192.168.2.1 255.255.255.0   // 配置接口 IP 地址
R1_config_f0/0#no shutdown         // 开启端口
R1_config_f0/0#^Z
R1#Jan  1 00:26:22 Configured from console 0 by


R1#show interface f0/0    // 查看端口信息
FastEthernet0/0 is up, line protocol is up
address is 00e0.0f26.0381
  MTU 1500 bytes, BW 100000 kbit, DLY 10 usec
  Interface address is 192.168.2.1/24
  Encapsulation ARPA
  ARP type: ARPA, ARP timeout 00:03:00
  60 second input rate 316 bits/sec, 0 packets/sec!
  60 second output rate 327 bits/sec, 0 packets/sec!
  Full-duplex, 100Mb/s, 100BaseTX, 508 ii, 808 oi
    510 packets input, 38501 bytes, 200 rx_freebuf
    Received 494 unicasts, 0 lowmark, 508 ri, 0 input errors
    0 overrun, 0 CRC, 0 framing, 0 busy, 0 long, 0 discard, 0 throttles
    471 packets output, 35453 bytes, 210 tx_freebd, 0 output errors
    0 underrun, 0 collisions, 0 late collisions, 0 deferred, 0 reTx expired
    0 resets, 0 lost carrier, 0 no carrier 0 grace stop 0 bus error
    0 output buffer failures, 0 output buffers swapped out 0 tx errors
    0 packets fast-switch
R1r#ping 192.168.2.10     // 检测与 PC 的连通性
PING 192.168.2.10 (192.168.2.10): 56 data bytes
!!!!!
--- 192.168.2.10 ping statistics ---
5 packets transmitted, 5 packets received, 0% packet loss
round-trip min/avg/max = 2/5/10ms


R1#write       // 保存配置文件
Saving current configuration...
OK!
```

3）查看路由器文件，并将配置文件下载到 TFTP 服务器上，代码如下。

```
R1#dir        // 查看路由器文件
Directory of /:
0   DCR26V1.3.3H.bin    <FILE>    5521346   Thu Jan  1 00:00:00 1970
```

```
    1  Function.map           <FILE>    792462   Thu Jan  1 00:00:00 1970
    2  startup-config         <FILE>       774   Tue Jan  1 00:30:44 2002
//系统默认的配置文件
free space 10420224

Router#copy startup-config tftp        //将配置文件复制到 TFTP 服务器
Remote-server ip address[]?192.168.2.10    //选择 TFTP 服务器地址
Destination file name[startup-config]?       //默认文件名（注：这部分最好用备份时间来对文件重命名，
```
命令为：**Destination file name[startup-config]?2012.7.31**）
```
#
TFTP:successfully send 2 blocks ,762 bytes    //备份成功
```

4）打开 TFTP 目录，使用写字板打开下载后的配置文件，修改机器名，再次使用 copy 命令上传到路由器中，重新启动后通过 show 命令观察到机器名已经被修改，代码如下。

```
R1#copy tftp:2012.7.31 startup-config   //从 TFTP 服务器复制文件 2012.7.31，并重命名为 startup-config
Remote-server ip address[]?192.168.2.10   //TFTP 服务器地址
#
       TFTP:successfully receive 2 blocks ,774 bytes    //复制文件成功

R1#reboot      //重新启动路由器
```

步骤 2：FTP 方式（采用 TCP，适合远程操作）。

1）设置 PC 网卡地址为 192.168.2.10，并安装 FTP 服务器。

2）设置 DCR-2655/2659/2855 的 F0/0 接口地址为 192.168.2.1，并测试连通性，代码如下。

```
Router>enable           //进入特权模式
Router#config           //进入全局模式
Router_config#hostname R1    //修改主机名为 R1
R1_config_f0/0#ip address 192.168.2.1 255.255.255.0   //配置接口 IP 地址
R1_config_f0/0#no shutdown    //开启端口
R1_config_f0/0#^Z
R1#Jan  1 00:26:22 Configured from console 0 by

R1#show interface f0/0     //查看端口信息
FastEthernet0/0 is up, line protocol is up
address is 00e0.0f26.0381
  MTU 1500 bytes, BW 100000 kbit, DLY 10 usec
  Interface address is 192.168.2.1/24
  Encapsulation ARPA
  ARP type: ARPA, ARP timeout 00:03:00
  60 second input rate 316 bits/sec, 0 packets/sec!
```

```
    60 second output rate 327 bits/sec, 0 packets/sec!
    Full-duplex, 100Mb/s, 100BaseTX, 508 ii, 808 oi
      510 packets input, 38501 bytes, 200 rx_freebuf
      Received 494 unicasts, 0 lowmark, 508 ri, 0 input errors
      0 overrun, 0 CRC, 0 framing, 0 busy, 0 long, 0 discard, 0 throttles
      471 packets output, 35453 bytes, 210 tx_freebd, 0 output errors
      0 underrun, 0 collisions, 0 late collisions, 0 deferred, 0 reTx expired
      0 resets, 0 lost carrier, 0 no carrier 0 grace stop 0 bus error
      0 output buffer failures, 0 output buffers swapped out 0 tx errors
      0 packets fast-switch
R1r#ping 192.168.2.10      // 检测与 PC 的连通性
PING 192.168.2.10 (192.168.2.10): 56 data bytes
!!!!!
--- 192.168.2.10 ping statistics ---
5 packets transmitted, 5 packets received, 0% packet loss
round-trip min/avg/max = 2/5/10ms


R1#write        // 保存配置文件
Saving current configuration...
OK!
```

3）查看路由器文件，并将配置文件下载到 FTP 服务器上，代码如下。

```
R1#dir        // 查看路由器文件
Directory of /:
0   DCR26V1.3.3H.bin    <FILE>    5521346   Thu Jan  1 00:00:00 1970
1   Function.map        <FILE>     792462   Thu Jan  1 00:00:00 1970
2   startup-config      <FILE>        774   Tue Jan  1 00:30:44 2002
// 系统默认的启动文件
free space 10420224


Router#copy startup-config ftp        // 将配置文件复制到 FTP 服务器
ftp user name[anonymous]?dcnu         // 输入 FTP 用户名
ftp user password[anonymous]?123456   // 输入 FTP 用户密码
Remote-server ip address[]?192.168.2.10   //FTP 服务器地址
Destination file name[startup-config]?    // 默认文件名
#
FTP:successfully send 2 blocks ,762 bytes    // 备份成功
```

4）打开 FTP 目录，使用写字板打开下载后的配置文件，修改机器名，再次使用 copy 命令上传到路由器中，重新启动后通过 show 命令观察到机器名已经被修改，代码如下。

```
R1#copy ftp startup-config 192.168.2.10    // 从 FTP 服务器复制文件
ftp user name[anonymous]?dcnu
ftp user password[anonymous]?123456
Source file name[]?
#
FTP:successfully receive 2 blocks ,774 bytes    // 复制文件成功


R1#reboot        // 重新启动路由器
```

步骤 3：启动到 monitor 模式。

当路由器的软件被破坏而无法启动时，可以在启动过程中按<Ctrl+Break>组合键，启动到 monitor 模式中，使用 ZMODEM 方式恢复文件。所谓 ZMODEM 方式是从路由器的 Console 端口以波特率规定的速率通过 PC 的串口传输文件的一种方式，不需要网线，但速度很慢。本实训以 TFTP 作为传输协议。

1）将路由器重启，在启动过程中按<Ctrl+Break>组合键，启动到 monitor 模式，代码如下。

```
System Bootstrap, Version 0.4.2
Serial num:8IRTJ610B407000041, ID num:200213
Copyright 2011 by Digital China Networks(BeiJing) Limited
Digital China-DCR-2659 Series DCR-2659
The current time: 65024-118-60 2:227:120

      Welcome to Digital China Multi-Protocol DCR-2659 Series Router
monitor#              // 进入 monitor 模式
```

2）设置 IP 地址，测试连通性，代码如下。

```
monitor#ip address 192.168.2.1 255.255.255.0    // 配置 IP 地址
monitor#ping 192.168.2.10            // 测试连通性
Ping 192.168.2.10 with 48 bytes of data:
Reply from 192.168.2.10: bytes=48 time=10ms TTL=128
Reply from 192.168.2.10: bytes=48 time=10ms TTL=128
Reply from 192.168.2.10: bytes=48 time=10ms TTL=128
Reply from 192.168.2.10: bytes=48 time=10ms TTL=128
4 packets sent, 4 packets received
round trip min/avg/max = 10/10/10 ms

monitor#
```

3）PC 启动 TFTP 服务，开始 copy 过程，代码如下。

```
monitor#copy tftp flash: 192.168.2.10    // 从 TFTP 服务器复制文件
Source file name[]?startup-config      // 需要复制的文件名
Destination file name[startup-config]?    // 默认文件名
#
```

TFTP:successfully receive 2 blocks ,778 bytes

monitor#

4) 重新启动设备，代码如下。

monitor#*reboot*

Do you want to reboot the router(y/n)?*y*

Please wait..

System Bootstrap, Version 0.4.2

Serial num:8IRTJ610B407000041, ID num:200213

Copyright 2011 by Digital China Networks(BeiJing) Limited

Digital China-DCR-2659 Series DCR-2659

The current time: 65024-118-60 2:227:120

Loading DCR26V1.3.3H.bin......

Start Decompress DCR26V1.3.3H.bin

##

##

##

Decompress 5519234 byte,Please wait system up..

Digital China Networks Limited Internetwork Operating System Software

Digital China-DCR-2659 Series Software , Version 1.3.3H, RELEASE SOFTWARE

System start up OK

Router console 0 is now available

Press RETURN to get started

   Welcome to DCR Multi-Protocol 1700 Series Router

Router>

步骤 4：清除密码。

当密码遗忘时，可以进入 monitor 模式，执行以下操作清除密码。

monitor#nopasswd

步骤 5：使用 TFTP 服务器维护路由器的原系统备份并升级成新版本。

1) 使用 show version 命令查看当前系统版本，代码如下。

Router#*show version*   // 查看路由器系统信息

Digital China Networks Limited Internetwork Operating System Software

DCR-2659 Series Software, Version 1.3.3H (MIDDLE), RELEASE SOFTWARE

Copyright 2011 by Digital China Networks(BeiJing) Limited

Compiled: 2011-01-19 15:23:40 by system, Image text-base: 0x6004

ROM: System Bootstrap, Version 0.4.2   // 系统引导版本信息

```
Serial num:8IRTJ610B407000041, ID num:200213
System image file is "DCR26V1.3.3H.bin"        // 配置文件
Digital China-DCR-2659 (PowerPC) Processor     // 路由器型号
65536K bytes of memory,16384K bytes of flash   // 内存容量
Router uptime is 0:00:00:57, The current time: 2002-01-01 00:00:57
Slot 0: SCC Slot
Port 0: 10/100Mbps full-duplex Ethernet     // 百兆全双工以太网口
Port 1: 2M full-duplex Serial                //2 Mbit/s 全双工串口
Port 2: 2M full-duplex Serial
Port 3: 1000Mbps full-duplex Ethernet       // 千兆全双工以太网口
Port 4: 1000Mbps full-duplex Ethernet
Port 5: 1000Mbps full-duplex Ethernet
Port 6: 1000Mbps full-duplex Ethernet
Router#
```

2）使用 dir 命令查看当前系统文件名称，代码如下。

```
Router#dir             // 使用 dir 命令查看系统文件
Directory of /:
0   DCR26V1.3.3H.bin    <FILE>    5521346   Thu Jan  1 00:00:00 1970
1   Function.map        <FILE>    792462    Thu Jan  1 00:00:00 1970
2   startup-config      <FILE>    778       Thu Jan  1 00:00:00 1970
free space 10420224
Router#
```

3）确保 TFTP 服务器与路由器连通，代码如下。

```
Router#ping 192.168.2.10
PING 192.168.2.10 (192.168.2.10): 56 data bytes
!!!!!
--- 192.168.2.10 ping statistics ---
5 packets transmitted, 5 packets received, 0% packet loss
round-trip min/avg/max = 0/0/0 ms
Router#
```

4）开启 TFTP 服务器，使用 copy file tftp 命令备份系统文件，代码如下。

```
Router#copy flash:DCR26V1.3.3H.bin tftp    // 将配置文件备份到 TFTP 服务器
Remote-server ip address[]?192.168.2.10    //TFTP 服务器地址
Destination file name[DCR26V1.3.3H.bin]?   // 默认文件名
################################################################
################################################################
################################################################
################################################################
```

```
###############################################################
###############################################################
###############################################################
###############################################################
###############################################################
TFTP:successfully send 10784 blocks ,5521346 bytes
Router#
```

5）在 TFTP 服务器中查看保存的文件，打开 TFTP 服务器的主目录，将看到文件，如图 5-6 所示。

图 5-6

6）在神州数码网站下载路由器新版本到 TFTP 服务器主目录中。

下载地址是 http://www.dcnetworks.com.cn/cn/download/，找到合适的设备型号开始下载，将下载文件保存到 TFTP 服务器的主目录中。

7）使用 copy tftp file 命令升级新版本，代码如下。

```
Router#copy tftp flash:                        // 从 TFTP 服务器复制文件
Source file name[]?DCR26V1.3.3H.bin            // 目标文件名（此处文件名只为举例，请根据实际情况输入）
Remote-server ip address[]?192.168.2.10        //TFTP 服务器地址
Destination file name[DCR26V1.3.3H.bin]?       // 默认文件名
###############################################################
###############################################################
###############################################################
###############################################################
###############################################################
###############################################################
###############################################################
###############################################################
###############################################################
###############################################################
TFTP:successfully receive 10784 blocks ,5521346 bytes
Router#
```

注意事项和排错

1）路由器以太网口和 PC 直接相连的时候，使用交叉双绞线。
2）关闭 PC 上的防火墙。
3）在实际工作中，通常使用日期或功能等标明配置文件。
4）使用 ZMODEM 方式，当文件比较大的时候，比较耗时。

共同思考

1) 请问 TFTP 和 FTP 这两种方式有什么区别？
2) 在 monitor 模式下，为什么不能使用 TFTP 方式？

实训 3　路由器直连路由的配置

实训目标

小刘通过之前的学习，已经掌握了企业级路由器的基本管理方法，现在他准备进入实际配置阶段，于是，他开始学习直连路由的配置。

通过学习本实训，能够掌握直连路由的概念，具备配置路由器直连路由的能力。

实训设备

1) DCR2655/2659/2855 路由器 1 台。
2) 交叉双绞线 2 条，配置线 1 条。
3) PC 2 台。

实训拓扑

实训拓扑图如图 5-7 所示。

图　5-7

实训任务

步骤 1：小刘先分别将 PC1 和 PC2 的以太网口接到了路由器的 G0/3 和 G0/4 端口，然后配置 PC 的 IP 地址，见表 5-2。

表　5-2

主　机　名	IP 地址	子 网 掩 码
PC1	192.168.3.10	255.255.255.0
PC2	192.168.4.10	255.255.255.0

步骤2：PC 互 ping，检测连通性，如图 5-8 和图 5-9 所示。

图 5-8

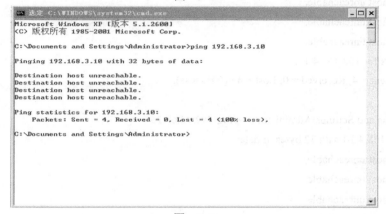

图 5-9

此时，小刘发现两台 PC 并不能互通，经过分析，他发现路由器的接口地址还没有配置，于是小刘开始着手配置路由器接口地址。

步骤3：配置路由器接口地址，并检测互通性，代码如下。

Router#*config*

Router_config#*interface G0/3*

Router_config_g0/3#*ip address 192.168.3.1 255.255.255.0*

Router_config_g0/3#*exit*

Router_config#*interface G0/4*

Router_config_g0/4#*ip address 192.168.4.1 255.255.255.0*

Router_config_g0/4#*exit*

Router_config#*exit*

Router#*ping 192.168.3.10*

PING 192.168.3.10 (192.168.3.10): 56 data bytes

!!!!!

--- 192.168.3.10 ping statistics ---

```
5 packets transmitted, 5 packets received, 0% packet loss
round-trip min/avg/max = 0/0/0 ms
Router#ping 192.168.4.10
PING 192.168.4.10 (192.168.4.10): 56 data bytes
!!!!!
--- 192.168.4.10 ping statistics ---
5 packets transmitted, 5 packets received, 0% packet loss
round-trip min/avg/max = 0/0/0 ms
```

步骤4：PC 互 ping，检测连通性，代码如下。

```
C:\Documents and Settings\Administrator>ping 192.168.4.10
Pinging 192.168.4.10 with 32 bytes of data:
Destination host unreachable.
Destination host unreachable.
Destination host unreachable.
Destination host unreachable.
Ping statistics for 192.168.4.10:
    Packets: Sent = 4, Received = 0, Lost = 4 (100% loss),

C:\Documents and Settings\Administrator>ping 192.168.3.10
Pinging 192.168.4.10 with 32 bytes of data:
Destination host unreachable.
Destination host unreachable.
Destination host unreachable.
Destination host unreachable.
Ping statistics for 192.168.3.10:
    Packets: Sent = 4, Received = 0, Lost = 4 (100% loss),
```

此时，小刘发现两台 PC 还是不能互相通信，经过仔细检查，他发现虽然已经配置了 PC 以及路由器接口的 IP 地址，但没有给 PC 配置网关，于是他按照表 5-3 又重新为 PC 配置了 IP，并重新检测连通性。

表 5-3

主 机 名	IP 地址	子 网 掩 码	网 关
PC1	192.168.3.10	255.255.255.0	192.168.3.1
PC2	192.168.4.10	255.255.255.0	192.168.4.1

检测连通性，代码如下。

```
C:\Documents and Settings\Administrator>ping 192.168.4.10
Pinging 192.168.3.10 with 32 bytes of data:
Reply from 192.168.3.10: bytes=32 time<1ms TTL=64
Reply from 192.168.3.10: bytes=32 time<1ms TTL=64
Reply from 192.168.3.10: bytes=32 time<1ms TTL=64
```

Reply from 192.168.3.10: bytes=32 time<1ms TTL=64
Ping statistics for 192.168.3.10:
    Packets: Sent = 4, Received = 4, Lost = 0 (0% loss),
Approximate round trip times in milli-seconds:
    Minimum = 0ms, Maximum = 0ms, Average = 0ms

C:\Documents and Settings\Administrator>*ping 192.168.3.10*
Pinging 192.168.3.10 with 32 bytes of data:
Reply from 192.168.3.10: bytes=32 time<1ms TTL=64
Reply from 192.168.3.10: bytes=32 time<1ms TTL=64
Reply from 192.168.3.10: bytes=32 time<1ms TTL=64
Reply from 192.168.3.10: bytes=32 time<1ms TTL=64
Ping statistics for 192.168.3.10:
    Packets: Sent = 4, Received = 4, Lost = 0 (0% loss),
Approximate round trip times in milli-seconds:
    Minimum = 0ms, Maximum = 0ms, Average = 0ms

注意事项和排错

1) 本实训中并未对中心路由器配置任何路由，而它却连通了两侧的网络段，并且只需要在两侧设备中配置到达非直连网络的默认路由，即可完成连通性。

2) 本实训的结论是：对路由器而言是否能够将数据转发到正确的地点，完全取决于其路由表的覆盖面，只有路由表中存在的表项才可以将数据按照路由表的指示传递出去，而对于路由表中没有标明出口的网络，路由器处于未知状态，对于目的地址是未知网络的数据，路由器将全部丢弃。

共同思考

该实训中 PC1 到 PC2 的报文转发流程是怎样的？

实训 4　路由器单臂路由的配置

实训目标

路由器的以太网端口通常用来连接企业的局域网，在很多时候，内网又划分了多个 VLAN，这些 VLAN 的用户都需要从一个出口访问外网。这个统一的出口在交换机中以 IEEE 802.1Q 的方式封装，意味着数据从这个出口访问对端设备时，必须能够识别并区分对待来自多个 VLAN 的数据才可以保证链路的正常通信。路由器以太网口在此时应该如何配置呢？这就是本实训需要解决的问题。

通过学习本实训，能够掌握单臂路由的概念，具备配置路由器单臂路由的能力。

实训设备

1) DCS 二层交换机 1 台。
2) DCR2655/2659/2855 路由器 1 台。
3) 直通双绞线 3 根，配置线 2 根。
4) PC 2 台。

实训拓扑

实训拓扑图如图 5-10 所示。

图 5-10

实训任务

步骤 1：配置交换机的 VLAN 及其成员端口，设置 24 端口的 Trunk 属性，配置 PVID 等，代码如下。

```
switch#
switch#config
switch(config)#vlan 10       // 划分 VLAN 10
switch(config-vlan10)#switchport interface e0/0/1-4
                             // 将 E0/0/1、E0/0/2、E0/0/3、E0/0/4 四个端口加入到 VLAN10 中
Set the port Ethernet0/0/1 access vlan 10 successfully
Set the port Ethernet0/0/2 access vlan 10 successfully
Set the port Ethernet0/0/3 access vlan 10 successfully
Set the port Ethernet0/0/4 access vlan 10 successfully
switch(config-vlan10)#exit
switch(config)#vlan 20   // 划分 VLAN20
switch(config-vlan20)#switchport interface e0/0/5-8
                             // 将 E0/0/5、E0/0/6、E0/0/7、E0/0/8 四个端口加入到 VLAN20 中
Set the port Ethernet0/0/5 access vlan 20 successfully
Set the port Ethernet0/0/6 access vlan 20 successfully
Set the port Ethernet0/0/7 access vlan 20 successfully
Set the port Ethernet0/0/8 access vlan 20 successfully
switch(config-vlan20)#exit
```

```
switch(config)#interface e0/0/24    // 进入 E0/0/24 端口
switch(config-if-ethernet0/0/24)#switchport mode trunk    // 将端口设为 Trunk 模式
Set the port Ethernet0/0/24 mode Trunk successfully
switch(config-if-ethernet0/0/24)#switchport trunk allowed vlan all    // 允许所有 VLAN 通过
set the Trunk port Ethernet0/0/24 allowed vlan successfully
switch(config-if-ethernet0/0/24)#
```

步骤 2：为路由器创建以太网接口的子接口，并在子接口上配置 VID 和对应的 IP 地址等，代码如下。

```
Router_config#interface f0/0.1    // 进入路由器子接口
Router_config_f0/0.1#ip address 192.168.1.1 255.255.255.0    // 为子接口添加 IP
Router_config_f0/0.1#encapsulation dot1Q 10    // 为 VLAN10 的数据包封装 802.1Q 协议
Router_config_f0/0.1#exit
Router_config#interface f0/0.2    // 进入路由器子接口
Router_config_f0/0.2#ip address 192.168.2.1 255.255.255.0    // 为子接口添加 IP
Router_config_f0/0.2#encapsulation dot1Q 20    // 为 VLAN20 的数据包封装 802.1Q 协议
Router_config_f0/0.2#exit
Router_config#
```

步骤 3：连接 PC，配置默认网关为路由器对应 VLAN 的接口地址，测试连通性。分别为 PC1 和 PC2 配置 IP 地址，并互 ping，见表 5-4。

表 5-4

主 机 名	IP 地址	子网掩码	默认网关
PC1	192.168.1.10	255.255.255.0	192.168.1.1
PC2	192.168.2.10	255.255.255.0	192.168.2.1

代码如下。

```
C:\Documents and Settings\Administrator>ping 192.168.2.10

Pinging 192.168.2.10 with 32 bytes of data:

Reply from 192.168.2.10: bytes=32 time<1ms TTL=63
Reply from 192.168.2.10: bytes=32 time<1ms TTL=63
Reply from 192.168.2.10: bytes=32 time<1ms TTL=63
Reply from 192.168.2.10: bytes=32 time<1ms TTL=63

Ping statistics for 192.168.2.10:
    Packets: Sent = 4, Received = 4, Lost = 0 (0% loss),
Approximate round trip times in milli-seconds:
    Minimum = 0ms, Maximum = 0ms, Average = 0ms

C:\Documents and Settings\Administrator>ping 192.168.1.10

Pinging 192.168.1.10 with 32 bytes of data:
```

Reply from 192.168.1.10: bytes=32 time<1ms TTL=63
Reply from 192.168.1.10: bytes=32 time<1ms TTL=63
Reply from 192.168.1.10: bytes=32 time<1ms TTL=63
Reply from 192.168.1.10: bytes=32 time<1ms TTL=63
Ping statistics for 192.168.1.10:
    Packets: Sent = 4, Received = 4, Lost = 0 (0% loss),
Approximate round trip times in milli-seconds:
    Minimum = 0ms, Maximum = 0ms, Average = 0ms

相关命令详解

相关命令详解，见表 5-5。

表 5-5

命 令	解 释
switchport interface ethernet 0/0/1-4	把端口 0/0/1 至 0/0/4 都加入指定 VLAN
switchport trunk vlan-allowed all	指定的 Trunk 口，所有 VLAN 可以通过

第 6 章　路由器路由技术基础

实训 1　路由器静态路由的配置

实训目标

路由技术与应用

小杜是一家集成公司的网络管理员，最近公司在本市又租下了一个办公地点，公司行政部门的同事已经把装修的事情处理差不多了，目前新办公地点和旧地点之间的网络还相互独立，他们希望能尽快让两部分的网络连通，就好启动新的办公地点了。小杜盘点了一下现有设备，发现可用的就是两台旧路由器，他又问了一下目前的上网情况，发现新大楼和旧楼都是同一个开发商的产业，因此互联网也都是通过开发商提供的宽带入口接入的。小杜决定把旧路由器用上，两边使用静态路由完成互联互通。经过师傅同意后，小杜开始工作了。

通过学习本实训，能够掌握静态路由的概念，具备配置静态路由的能力。

实训设备

1）DCR2655/2659/2855 路由器 3 台。
2）CR-V35FC/CR-V35MT 各 1 条。
3）交叉双绞线 3 条。
4）PC 1 台。

实训拓扑

实训拓扑图如图 6-1 所示。

图 6-1

实训任务

步骤 1：按照表 6-1 配置所有接口的 IP 地址，保证所有接口全部是 up 状态，测试连通性。

表 6-1

RouteA		RouteB		RouteC	
S0/2(DCE)	192.168.1.1	S0/1(DTE)	192.168.1.2	F0/0	192.168.2.2
F0/0	192.168.0.1	F0/0	192.168.2.1	G0/3	192.168.3.1

RouteA：

RouteA_config#

RouteA_config#*interface s0/2*　　　　　//进入 S0/2 端口

RouteA_config_s0/2#*ip address 192.168.1.1 255.255.255.0*　　//配置接口 IP 地址

RouteA_config_s0/2#*no shutdown*　　　　//开启端口

RouteA_config_s0/2#*physical-layer speed 64000*　　//配置时钟，V35 线的 DCE 端必须要配置时钟

RouteA_config_s0/2#Jan　1 00:01:08 Line on Interface Serial0/2, changed to up

Jan　1 00:01:09 Line protocol on Interface Serial0/2, change state to up

RouteA_config_s0/2#*exit*

RouteA_config#*interface f0/0*　　　　　//进入 f0/0 端口

RouteA_config_f0/0#*ip address 192.168.0.1 255.255.255.0*　　//配置对口 IP 地址

RouteA_config_f0/0#*no shutdown*　　//开启端口

RouteA_config_f0/0#*exit*

RouteB：

RouteB_config#

RouteB _config#*interface s0/1*

RouteB _config_s0/1#*ip address 192.168.1.2 255.255.255.0*

RouteB _config_s0/1#*no shutdown*

RouteB _config_s0/1#*exit*

RouteB _config#*interface f0/0*

RouteB _config_f0/0#*ip address 192.168.2.1 255.255.255.0*

RouteB _config_f0/0#*no shutdown*

RouteB _config_f0/0#*exit*

RouteC：

RouteC_config#

RouteC_config#*interface f0/0*

RouteC_config_f0/0#*ip address 192.168.2.2 255.255.255.0*

RouteC_config_f0/0#*no shutdown*

RouteC_config_f0/0#*exit*

RouteC_config#*interface g0/3*

RouteC_config_g0/3#*ip address 192.168.3.1 255.255.255.0*

RouteC_config_g0/3#*no shutdown*

RouteC_config_g0/3#*exit*

RouteC_config#

步骤2：查看路由器A的路由表，代码如下。

RouteA#*show ip route*

Codes: C - connected, S - static, R - RIP, B - BGP, BC - BGP connected

       D - BEIGRP, DEX - external BEIGRP, O - OSPF, OIA - OSPF inter area

       ON1 - OSPF NSSA external type 1, ON2 - OSPF NSSA external type 2

       OE1 - OSPF external type 1, OE2 - OSPF external type 2

       DHCP - DHCP type, L1 - IS-IS level-1, L2 - IS-IS level-2

VRF ID: 0

C   192.168.0.0/24   is directly connected, FastEthernet0/0

C   192.168.1.0/24   is directly connected, Serial0/2

步骤3：查看路由器B的路由表，代码如下。

RouteB#*show ip route*

Codes: C - connected, S - static, R - RIP, B - BGP, BC - BGP connected

       D - BEIGRP, DEX - external BEIGRP, O - OSPF, OIA - OSPF inter area

       ON1 - OSPF NSSA external type 1, ON2 - OSPF NSSA external type 2

       OE1 - OSPF external type 1, OE2 - OSPF external type 2

       DHCP - DHCP type, L1 - IS-IS level-1, L2 - IS-IS level-2

VRF ID: 0

C   192.168.1.0/24   is directly connected, Serial0/1

C   192.168.2.0/24   is directly connected, FastEthernet0/0

步骤4：查看路由器C的路由表，代码如下。

RouteC#*show ip route*

Codes: C - connected, S - static, R - RIP, B - BGP, BC - BGP connected

       D - BEIGRP, DEX - external BEIGRP, O - OSPF, OIA - OSPF inter area

       ON1 - OSPF NSSA external type 1, ON2 - OSPF NSSA external type 2

       OE1 - OSPF external type 1, OE2 - OSPF external type 2

       DHCP - DHCP type, L1 - IS-IS level-1, L2 - IS-IS level-2

VRF ID: 0

C   192.168.2.0/24   is directly connected, FastEthernet0/0

C   192.168.3.0/24   is directly connected, GigaEthernet0/3

步骤5：在路由器A上ping路由器C，代码如下。

Router A#*ping 192.168.2.2*

PING 192.168.2.2 (192.168.2.2): 56 data bytes

……

--- 192.168.2.2 ping statistics ---
5 packets transmitted, 0 packets received, 100% packet loss   // 不通

步骤6：在路由器 A 上配置静态路由，代码如下。

Router A#*config*
Router A_config#*ip route 192.168.2.0 255.255.255.0 192.168.1.2*// 配置目标网段和下一跳
Router A_config#*ip route 192.168.3.0 255.255.255.0 192.168.1.2*

步骤7：查看路由表，代码如下。

RouteA#*show ip route*
Codes: C - connected, S - static, R - RIP, B - BGP, BC - BGP connected
       D - BEIGRP, DEX - external BEIGRP, O - OSPF, OIA - OSPF inter area
       ON1 - OSPF NSSA external type 1, ON2 - OSPF NSSA external type 2
       OE1 - OSPF external type 1, OE2 - OSPF external type 2
       DHCP - DHCP type, L1 - IS-IS level-1, L2 - IS-IS level-2
VRF ID: 0
C   192.168.0.0/24   is directly connected, FastEthernet0/0
C   192.168.1.0/24   is directly connected, Serial0/2
S   192.168.2.0/24   [1,0] via 192.168.1.2(on Serial0/2)
S   192.168.3.0/24   [1,0] via 192.168.1.2(on Serial0/2)

步骤8：配置路由器 B 的静态路由并查看路由表，代码如下。

RouteB#*config*
RouteB_config#*ip route 192.168.0.0 255.255.255.0 192.168.1.1*
RouteB_config#*ip route 192.168.3.0 255.255.255.0 192.168.1.1*
RouteB_config#*exit*
RouteB#Jan   1 00:22:04 Configured from console 0 by UNKNOWN
RouteB#*show ip route*
Codes: C - connected, S - static, R - RIP, B - BGP, BC - BGP connected
       D - BEIGRP, DEX - external BEIGRP, O - OSPF, OIA - OSPF inter area
       ON1 - OSPF NSSA external type 1, ON2 - OSPF NSSA external type 2
       OE1 - OSPF external type 1, OE2 - OSPF external type 2
       DHCP - DHCP type, L1 - IS-IS level-1, L2 - IS-IS level-2
VRF ID: 0
S   192.168.0.0/24   [1,0] via 192.168.1.1(on Serial0/1)
C   192.168.1.0/24   is directly connected, Serial0/1
C   192.168.2.0/24   is directly connected, FastEthernet0/0
S   192.168.3.0/24   [1,0] via 192.168.1.1(on Serial0/1)

步骤9：配置路由器 C 的静态路由并查看路由表，代码如下。

RouteC#*config*

RouteC_config#*ip route 192.168.0.0 255.255.0.0 192.168.2.1*   // 采用超网的方法
RouteC_config#^*Z*
RouteC#*show ip route*
Codes: C - connected, S - static, R - RIP, B - BGP
       D - DEIGRP, DEX - external DEIGRP, O - OSPF, OIA - OSPF inter area
       ON1 - OSPF NSSA external type 1, ON2 - OSPF NSSA external type 2
       OE1 - OSPF external type 1, OE2 - OSPF external type 2
S   192.168.0.0/16   [1,0] via 192.168.2.1(on FastEthernet0/0)   // 注意子网掩码为 16 位
C   192.168.2.0/24   is directly connected, FastEthernet0/0
C   192.168.3.0/24   is directly connected, GigaEthernet0/3

步骤 10：测试连通性，代码如下。

Router-C#*ping 192.168.0.1*
PING 192.168.0.1 (192.168.0.1): 56 data bytes
!!!!!   // 成功
--- 192.168.0.1 ping statistics ---
5 packets transmitted, 5 packets received, 0% packet loss
round-trip min/avg/max = 30/32/40 ms

注意事项和排错

1）非直连的网段都要配置路由器。
2）以太网接口要接主机或交换机才能 up。
3）串口注意 DCE 和 DTE 的问题。

相关命令详解

相关命令详解，见表 6-2。

表 6-2

命　　令	解　　释
ip route 192.168.2.0 255.255.255.0 192.168.1.2	静态路由，把 192.168.2.0/24 网段流入的数据，从下一跳为 192.168.1.2 的端口流出

共同思考

1）什么情况下可以采用路由器 C 的超网配置方法？
2）静态路由有什么优势？什么情况下使用？
3）路由器 B 如果不配置任何静态路由，会影响哪些网段间的互通？

实训 2　静态路由掩码最长匹配

实训目标

面对当前 IP 地址资源匮乏的情况，很多企业采用子网的方式设置局域网环境，如果在三层设备中为每一个子网提供一条确切的路由，当然是最标准的路由设置方法，但这样也会造成三层设备的路由表变大，因此查询效率受到一定的影响。比较合理的做法就是将可能合并成一个大网的子网在路由表中进行汇总，使它们通过一条路由查询到出口。这样做，查询效率会提升，但同时也要求必须对全网的子网分布作合理规划，否则会因为路由表的查询原则造成网络的连通问题。本实训集中讨论关于路由表掩码最长匹配原则，主要针对 VLSM 环境进行讨论。

通过学习本实训，能够掌握路由掩码的概念，具备配置静态路由掩码的能力。

实训设备

1）DCR2655/2659/2855 路由器 2 台。
2）交叉双绞线 3 条。
3）PC 2 台。

实训拓扑

实训拓扑图如图 6-2 所示。

图　6-2

实训任务

步骤 1：参照图 6-2 所示的拓扑图，按照表 6-3 配置接口。

表　6-3

主　机　名	F0/0	G0/3	PC 地址	默　认　网　关
RouteA	10.1.1.1	10.1.3.1	—	—
RouteB	10.1.1.2	10.1.2.1	—	—
PC1	—	—	10.1.3.10/24	10.1.3.1
PC2	—	—	10.1.2.10/24	10.1.2.1

第 6 章 路由器路由技术基础

RouteA:
```
Router#
Router#config                            //全局配置模式
Router_config#hostname RouteA            //更改主机名
RouteA_config#interface f 0/0            //进入 f0/0 端口
RouteA_config_f0/0#ip address 10.1.1.1 255.255.255.0   //配置接口地址
RouteA_config_f0/0#no shutdown           //开启端口
RouteA_config_f0/0#exit
RouteA_config#interface g 0/3
RouteA_config_g0/3#ip add 10.1.3.1 255.255.255.0
RouteA_config_g0/3#no shutdown
RouteA_config_g0/3#exit
RouteA_config#write                      //保存配置
Saving current configuration...
OK!
RouteA_config#
```

RouteB:
```
Router_config#hostname RouteB
RouteB_config#interface f 0/0
RouteB_config_f0/0#ip add 10.1.1.2 255.255.255.0
RouteB_config_f0/0#exit
RouteB_config#ping 10.1.1.1              //检测与路由器 A 的连通性
PING 10.1.1.1 (10.1.1.1): 56 data bytes
!!!!!
--- 10.1.1.1 ping statistics ---
5 packets transmitted, 5 packets received, 0% packet loss
round-trip min/avg/max = 0/0/0 ms
RouteB_config#interface g 0/3
RouteB_config_g0/3#ip add 10.1.2.1 255.255.255.0
RouteB_config_g0/3#no shutdown
RouteB_config_g0/3#exit
RouteB_config#wr
Saving current configuration...
OK!
RouteB_config#
```

注意：以上已经验证了两台路由器之间单条链路的连通性。

步骤 2：配置静态路由。

Route A:
```
RouteA_config#ip route 10.0.0.0  255.0.0.0  10.1.1.2       //配置静态路由
```

RouteA_config#*ip route 10.1.2.0 255.255.255.0 10.1.3.1*   // 配置静态路由

注意：此处指明的下一跳路由是错误的，目的是通过最长匹配原则让路由器将数据发往错误的地点从而引起不通。

Route B：

RouteB _config#*ip route 10.0.0.0 255.0.0.0 10.1.1.1*   // 配置静态路由

RouteB _config#*ip route 10.1.3.0 255.255.255.0 10.1.2.1*   // 配置静态路由

注意：此处指明的下一跳路由是错误的，目的是通过最长匹配原则让路由器将数据发往错误的地点从而引起不通。

步骤3：查看路由表。

RouteA：

RouteA#show ip route

Codes: C - connected, S - static, R - RIP, B - BGP, BC - BGP connected

      D - BEIGRP, DEX - external BEIGRP, O - OSPF, OIA - OSPF inter area

      ON1 - OSPF NSSA external type 1, ON2 - OSPF NSSA external type 2

      OE1 - OSPF external type 1, OE2 - OSPF external type 2

      DHCP - DHCP type, L1 - IS-IS level-1, L2 - IS-IS level-2

VRF ID: 0

S	10.0.0.0/8	[1,0] via 10.1.1.2(on FastEthernet0/0)
C	10.1.1.0/24	is directly connected, FastEthernet0/0
S	10.1.2.0/24	[1,0] via 10.1.3.1 (on GigaEthernet0/3)
C	10.1.3.0/24	is directly connected, GigaEthernet0/3

RouteB：

RouteB#show ip route

Codes: C - connected, S - static, R - RIP, B - BGP, BC - BGP connected

      D - BEIGRP, DEX - external BEIGRP, O - OSPF, OIA - OSPF inter area

      ON1 - OSPF NSSA external type 1, ON2 - OSPF NSSA external type 2

      OE1 - OSPF external type 1, OE2 - OSPF external type 2

      DHCP - DHCP type, L1 - IS-IS level-1, L2 - IS-IS level-2

VRF ID: 0

S	10.0.0.0/8	[1,0] via 10.1.1.1(on FastEthernet0/0)
C	10.1.1.0/24	is directly connected, FastEthernet0/0
C	10.1.2.0/24	is directly connected, GigaEthernet0/3
S	10.1.3.0/24	[1,0] via 10.1.2.10(on GigaEthernet0/3)

步骤4：测试连通性。

PC1：

C:\>*ipconfig*

Windows IP Configuration

```
Ethernet adapter 本地连接:
        Connection-specific DNS Suffix  . :
        IP Address. . . . . . . . . . . : 10.1.3.10
        Subnet Mask . . . . . . . . . . : 255.255.255.0
        Default Gateway . . . . . . . . : 10.1.3.1
```

```
C:\>ping 10.1.3.1
Pinging 10.1.3.1 with 32 bytes of data:
Reply from 10.1.2.1: bytes=32 time<1ms TTL=255
Reply from 10.1.2.1: bytes=32 time<1ms TTL=255
Reply from 10.1.2.1: bytes=32 time<1ms TTL=255
Reply from 10.1.2.1: bytes=32 time<1ms TTL=255
Ping statistics for 10.1.3.1:
    Packets: Sent = 4, Received = 4, Lost = 0 (0% loss),
Approximate round trip times in milli-seconds:
Minimum = 0ms, Maximum = 0ms, Average = 0ms
```

```
C:\>ping 10.1.2.1
Pinging 10.1.2.1 with 32 bytes of data:
Request timed out.
Request timed out.
Request timed out.
Request timed out.
Ping statistics for 10.1.2.1:
Packets: Sent = 4, Received = 0, Lost = 4 (100% loss),
```

此时看到 PC1 到 10.1.2.1 是连不通的,进一步跟踪路径得到:

```
C:\>tracert 10.1.2.10
Tracing route to 10.1.2.10 over a maximum of 30 hops

  1    <1 ms    <1 ms    <1 ms    10.1.3.1
  2     *        *        *       Request timed out.
  3    ^C
C:\>
```

此时查看到问题出在 10.1.2.1 的下一步无法到达。

注意:PC2 情况类似,此处略。

步骤 5:问题分析和解决。

从上一步了解到不能连通的原因是 10.1.3.1 的路由出了问题。因此回顾 RouteA 的路由表有如下的项:

S	10.1.2.0/24	[1,0] via 10.1.3.1 (on GigaEthernet 0/3)
S	10.0.0.0/8	[1,0] via 10.1.1.1(on FastEthernet0/0)

由于目标为 10.1.2.0 网络的成员，因此根据最长匹配的原则，这里从 G0/3 口发送出去，显然它找不到正确的位置了，要解决上述问题有两种办法：

① 将目标地址改为其他网段，比如 10.1.4.0，这样通过路由表，RouteA 会将数据从 f0/0 发送出去，因此可以到达，但这需要更改 RouteB 的 G0/3 接口地址为 10.1.4.1，PC2 地址也需做相应修改。

② 更改静态路由配置，具体修改方法如下：

----- RouteA ------ 删除 10.1.2.0 的路由

----- RouteB ------ 删除 10.1.3.0 的路由

在设备中实现如下：

RouteA：

RouteA_config#*no ip route 10.1.2.0 255.255.255.0 10.1.3.1*

RouteA _config#*sh ip route*

Codes: C - connected, S - static, R - RIP, B - BGP, BC - BGP connected

       D - DEIGRP, DEX - external DEIGRP, O - OSPF, OIA - OSPF inter area

       ON1 - OSPF NSSA external type 1, ON2 - OSPF NSSA external type 2

       OE1 - OSPF external type 1, OE2 - OSPF external type 2

       DHCP - DHCP type

VRF ID: 0

S	10.0.0.0/8	[1,0] via 10.1.1.1(on FastEthernet0/0)
C	10.1.1.0/24	is directly connected, FastEthernet0/0
C	10.1.2.0/24	is directly connected, GigaEthernet0/3

RouteB：

RouteB_config #*no ip route 10.1.3.0 255.255.255.0 10.1.2.1*

RouteB_config #*show ip route*

Codes: C - connected, S - static, R - RIP, B - BGP, BC - BGP connected

       D - DEIGRP, DEX - external DEIGRP, O - OSPF, OIA - OSPF inter area

       ON1 - OSPF NSSA external type 1, ON2 - OSPF NSSA external type 2

       OE1 - OSPF external type 1, OE2 - OSPF external type 2

       DHCP - DHCP type

VRF ID: 0

S	10.0.0.0/8	[1,0] via 10.1.1.1(on FastEthernet0/0)
C	10.1.1.0/24	is directly connected, FastEthernet0/0
C	10.1.2.0/24	is directly connected, GigaEthernet0/3

再次测试连通性：

C:\>*ping 10.1.3.10*

Pinging 10.1.3.10 with 32 bytes of data:

Reply from 10.1.3.10: bytes=32 time=2ms TTL=126

> Reply from 10.1.3.10: bytes=32 time<1ms TTL=126
> Reply from 10.1.3.10: bytes=32 time<1ms TTL=126
> Reply from 10.1.3.10: bytes=32 time<1ms TTL=126
> Ping statistics for 10.1.3.10:
>   Packets: Sent = 4, Received = 4, Lost = 0 (0% loss),
> Approximate round trip times in milli-seconds:
> Minimum = 0ms, Maximum = 2ms, Average = 0ms

此时已经可以连通。

注意事项和排错

实训时在步骤 1 将设备 hostname 改为与拓扑图一致的状态，有助于实训的顺利进行。

本实训最终设置的静态路由的掩码并非最优方案，只是出于本实训测试目的，请正常设置静态路由时不要完全效仿。

如果实训过程中发生静态路由无法写入路由表的情况，则查看静态路由指向的下一跳所对应端口是否处于 up 状态。

实训 3　路由器 RIP 的配置

RIP 路由协议

实训目标

小宋是一所学校的网络管理员，最近学校要进行一次网络改造，在规划网络时，小宋发现改造后的网络中路由器较多，手工配置静态路由会带来很大的工作负担，同时，小宋认为在不太稳定的网络环境里，手工修改路由表不现实，于是他考虑采用动态路由协议。本实训就帮助我们理解如何配置动态路由，同时理解路由协议的工作过程。

通过学习本实训，能够掌握 RIP 协议，具备配置 RIP 路由功能的能力。

实训设备

1）DCR2655/2659/2855 路由器 3 台。
2）CR-V35FC1/CR-V35MT1 各 1 条。
3）交叉双绞线 3 条。
4）PC 2 台。

实训拓扑

实训拓扑图如图 6-3 所示。

图 6-3

实训任务

步骤 1：按照表 6-4 配置所有接口的 IP 地址，保证所有接口全部是 up 状态，测试连通性。

表 6-4

RouteA		RouteB		RouteC	
S0/2(DCE)	192.168.1.1	S0/1 (DTE)	192.168.1.2	F0/0	192.168.2.2
F0/0	192.168.0.1	F0/0	192.168.2.1	G0/3	192.168.3.1

RouteA：

RouteA_config#

RouteA_config#*interface s0/2*   // 进入 S0/2 端口

RouteA_config_s0/2#*ip address 192.168.1.1 255.255.255.0*   // 配置接口 IP 地址

RouteA_config_s0/2#*no shutdown*   // 开启端口

RouteA_config_s0/2#*physical-layer speed 64000*   // 配置时钟，V35 线的 DCE 端必须要配置时钟

RouteA_config_s0/2#Jan 1 00:01:08 Line on Interface Serial0/2, changed to up

Jan 1 00:01:09 Line protocol on Interface Serial0/2, change state to up

RouteA_config_s0/2#*exit*

RouteA_config#*interface f0/0*   // 进入 f0/0 端口

RouteA_config_f0/0#*ip address 192.168.0.1 255.255.255.0*   // 配置端口地址

RouteA_config_f0/0#*no shutdown*   // 开启端口

RouteA_config_f0/0#*exit*

RouteB：

RouteB_config#

RouteB _config#*interface s0/1*

RouteB _config_s0/1#*ip address 192.168.1.2 255.255.255.0*

RouteB _config_s0/1#*no shutdown*

RouteB _config_s0/1#*exit*

RouteB _config#*interface f0/0*

RouteB _config_f0/0#*ip address 192.168.2.1 255.255.255.0*
RouteB _config_f0/0#*no shutdown*
RouteB _config_f0/0#*exit*

RouteC：

RouteC_config#
RouteC_config#*interface f0/0*
RouteC_config_f0/0#*ip address 192.168.2.2 255.255.255.0*
RouteC_config_f0/0#*no shutdown*
RouteC_config_f0/0#*exit*
RouteC_config#*interface g0/3*
RouteC_config_g0/3#*ip address 192.168.3.1 255.255.255.0*
RouteC_config_g0/3#*no shutdown*
RouteC_config_g0/3#*exit*
RouteC_config#

步骤 2：查看路由器 A 的路由表，代码如下。

RouteA#*show ip route*
Codes: C - connected, S - static, R - RIP, B - BGP, BC - BGP connected
　　　　D - BEIGRP, DEX - external BEIGRP, O - OSPF, OIA - OSPF inter area
　　　　ON1 - OSPF NSSA external type 1, ON2 - OSPF NSSA external type 2
　　　　OE1 - OSPF external type 1, OE2 - OSPF external type 2
　　　　DHCP - DHCP type, L1 - IS-IS level-1, L2 - IS-IS level-2
VRF ID: 0
C　　　 192.168.0.0/24　　　 is directly connected, FastEthernet0/0
C　　　 192.168.1.0/24　　　 is directly connected, Serial0/2

步骤 3：查看路由器 B 的路由表，代码如下。

RouteB#*show ip route*
Codes: C - connected, S - static, R - RIP, B - BGP, BC - BGP connected
　　　　D - BEIGRP, DEX - external BEIGRP, O - OSPF, OIA - OSPF inter area
　　　　ON1 - OSPF NSSA external type 1, ON2 - OSPF NSSA external type 2
　　　　OE1 - OSPF external type 1, OE2 - OSPF external type 2
　　　　DHCP - DHCP type, L1 - IS-IS level-1, L2 - IS-IS level-2
VRF ID: 0
C　　　 192.168.1.0/24　　　 is directly connected, Serial0/1
C　　　 192.168.2.0/24　　　 is directly connected, FastEthernet0/0

步骤 4：查看路由器 C 的路由表，代码如下。

RouteC#*show ip route*
Codes: C - connected, S - static, R - RIP, B - BGP, BC - BGP connected

```
            D - BEIGRP, DEX - external BEIGRP, O - OSPF, OIA - OSPF inter area
            ON1 - OSPF NSSA external type 1, ON2 - OSPF NSSA external type 2
            OE1 - OSPF external type 1, OE2 - OSPF external type 2
            DHCP - DHCP type, L1 - IS-IS level-1, L2 - IS-IS level-2
VRF ID: 0
C        192.168.2.0/24            is directly connected, FastEthernet0/0
C        192.168.3.0/24            is directly connected, GigaEthernet0/3
```

步骤 5：在路由器 A 上 ping 路由器 C，代码如下。

```
Router A#ping 192.168.2.2
PING 192.168.2.2 (192.168.2.2): 56 data bytes
.....
--- 192.168.2.2 ping statistics ---
5 packets transmitted, 0 packets received, 100% packet loss           // 不通
```

步骤 6：在路由器 A 上配置 RIP 并查看路由表，代码如下。

```
RouteA_config#
RouteA_config#router rip                    // 启动 RIP
RouteA_config_rip#network 192.168.0.0       // 宣告网段
RouteA_config_rip#network 192.168.1.0
RouteA_config_rip#^Z
RouteA#Jan  1 00:11:37 Configured from console 0 by
RouteA#show ip route
Codes: C - connected, S - static, R - RIP, B - BGP, BC - BGP connected
       D - BEIGRP, DEX - external BEIGRP, O - OSPF, OIA - OSPF inter area
       ON1 - OSPF NSSA external type 1, ON2 - OSPF NSSA external type 2
       OE1 - OSPF external type 1, OE2 - OSPF external type 2
       DHCP - DHCP type, L1 - IS-IS level-1, L2 - IS-IS level-2
VRF ID: 0
C        192.168.0.0/24            is directly connected, FastEthernet0/0
C        192.168.1.0/24            is directly connected, Serial0/2
```

注意：路由表中并没有出现 RIP 学习到的路由。

步骤 7：在路由器 B 上配置 RIP 并查看路由表，代码如下。

```
RouteB_config#router rip
RouteB_config_rip#network 192.168.1.0
RouteB_config_rip#network 192.168.2.0
RouteB_config_rip#^Z
RouteB#Jan  1 00:15:31 Configured from console 0 by
RouteB#
RouteB#show ip route
```

Codes: C - connected, S - static, R - RIP, B - BGP, BC - BGP connected
       D - BEIGRP, DEX - external BEIGRP, O - OSPF, OIA - OSPF inter area
       ON1 - OSPF NSSA external type 1, ON2 - OSPF NSSA external type 2
       OE1 - OSPF external type 1, OE2 - OSPF external type 2
       DHCP - DHCP type, L1 - IS-IS level-1, L2 - IS-IS level-2
VRF ID: 0
R   192.168.0.0/24   [120,1] via 192.168.1.1(on Serial0/1) // 从路由器 A 学习到的路由
C   192.168.1.0/24   is directly connected, Serial0/1
C   192.168.2.0/24   is directly connected, FastEthernet0/0
RouteB#

步骤 8：在路由器 C 上配置 RIP 并查看路由表，代码如下。

Router-C_config#***router rip***
Router-C_config_rip#***network 192.168.2.0***
Router-C_config_rip#***network 192.168.3.0***
Router-C_config_rip#^Z
RouteC#***show ip route***
Codes: C - connected, S - static, R - RIP, B - BGP, BC - BGP connected
       D - BEIGRP, DEX - external BEIGRP, O - OSPF, OIA - OSPF inter area
       ON1 - OSPF NSSA external type 1, ON2 - OSPF NSSA external type 2
       OE1 - OSPF external type 1, OE2 - OSPF external type 2
       DHCP - DHCP type, L1 - IS-IS level-1, L2 - IS-IS level-2
VRF ID: 0
R   192.168.1.0/24   [120,1] via 192.168.2.1(on FastEthernet0/0)
C   192.168.2.0/24   is directly connected, FastEthernet0/0
C   192.168.3.0/24   is directly connected, GigaEthernet0/3

步骤 9：再次查看 A 和 B 的路由表，代码如下。

RouteA#***show ip route***
Codes: C - connected, S - static, R - RIP, B - BGP, BC - BGP connected
       D - BEIGRP, DEX - external BEIGRP, O - OSPF, OIA - OSPF inter area
       ON1 - OSPF NSSA external type 1, ON2 - OSPF NSSA external type 2
       OE1 - OSPF external type 1, OE2 - OSPF external type 2
       DHCP - DHCP type, L1 - IS-IS level-1, L2 - IS-IS level-2
VRF ID: 0
C   192.168.1.0/24   is directly connected, Serial0/2
R   192.168.2.0/24   [120,1] via 192.168.1.2(on Serial0/2)
R   192.168.3.0/24   [120,2] via 192.168.1.2(on Serial0/2)

RouteB#***show ip route***
Codes: C - connected, S - static, R - RIP, B - BGP, BC - BGP connected

```
            D - BEIGRP, DEX - external BEIGRP, O - OSPF, OIA - OSPF inter area
            ON1 - OSPF NSSA external type 1, ON2 - OSPF NSSA external type 2
            OE1 - OSPF external type 1, OE2 - OSPF external type 2
            DHCP - DHCP type, L1 - IS-IS level-1, L2 - IS-IS level-2
VRF ID: 0
C       192.168.1.0/24          is directly connected, Serial0/1
C       192.168.2.0/24          is directly connected, FastEthernet0/0
R       192.168.3.0/24          [120,1] via 192.168.2.2(on FastEthernet0/0)
```

注意：所有网段都学习到了路由。

步骤10：相关的查看命令，代码如下。

```
RouteA#sh ip rip                // 显示 RIP 状态
Update: 30, Expire: 180, Holddown: 120
Input-queue: 50
Validate-update-source: Enable
Neighbor List:


RouteA#show ip rip protocol     // 显示协议细节
RIP is Active
Update interval 30(s), Invalid interval 180(s)      // 注意定时器的值
Holddown interval 120(s), Trigger interval 5(s)
Automatic network summarization: Enable
Network List:
   network 192.168.0.0
   network 192.168.1.0
Filter list:
Offset list:
Redistribute policy:
Interface send version and receive version:
Global version : default
   Interface            Send-version    Recv-version    Nbr_number
   Serial0/2            V1              V1 V2           2
   FastEthernet0/0      V1              V1 V2           0
Distance: 0 (default is 120):        // 注意默认的管理距离
Maximum route count: 1024,    Current route count:4


RouteA#show ip rip database          // 显示 RIP 数据库
192.168.0.0/24       directly connected    FastEthernet0/0
192.168.1.0/24       directly connected    Serial0/2
```

```
    192.168.2.0/24      [120,1]    via 192.168.1.2 (on Serial0/2)     00:00:04      // 收到 RIP 广播的时间
    192.168.3.0/24      [120,16]   via 192.168.1.2 holddown (on Serial0/2)  00:00:10
RouteA#show ip route rip              // 仅显示 RIP 学习到的路由
R   192.168.2.0/24      [120,1] via 192.168.1.2(on Serial0/2)
R   192.168.3.0/24      [120,2] via 192.168.1.2(on Serial0/2)
```

相关命令详解

在 DCNOS 中运行 RIP 的基本配置很简单，通常只需打开 RIP 开关，并且配置运行 RIP 的网段，即按 RIP 默认配置发送和接收 RIP 数据报。如果需要则可以切换发送、接收 RIP 数据报的版本，允许/禁止发送、接收 RIP 数据报，见表 6-5。

表 6-5

命令	解释
router rip	打开 RIP
no router rip	关闭 RIP
network <A.B.C.D/M | ifname>	设定运行 RIP 的网段
no network <A.B.C.D/M | ifname>	删除运行 RIP 的网段

共同思考

1）为什么路由器 B 没有配置 RIP 时，路由器 A 没有出现 RIP 路由？
2）如果不是连续的子网，那么会出现什么结果？
3）RIP 的广播周期是多少？

实训 4　路由器单区域 OSPF 协议的配置

实训目标

OSPF 路由协议

OSPF 协议具有快速收敛、无自环、区域划分、等价路由、路由分级、支持验证、组播发送等特性，在大规模网络中，OSPF 作为链路状态路由协议的代表应用非常广泛。通过本实训，可以掌握单区域 OSPF 的配置，从而更好地理解链路状态路由协议的工作过程。

通过学习本实训，能够掌握 OSPF 协议，具备配置 OSPF 路由功能的能力。

实训设备

1）DCR2655/2659/2855 路由器 2 台。
2）背靠背线 1 条。

实训拓扑

实训拓扑图如图 6-4 所示。

图 6-4

实训任务

IP 地址表，见表 6-6。

表 6-6

设备	S0/1(DTE)	S0/2(DCE)	Loopback
RouteA	—	192.168.1.1/24	10.10.10.1/24
RouteB	192.168.1.2/24	—	10.10.11.1/24

步骤 1：路由器环回接口的配置，代码如下。

路由器 A：

```
RouteA_config#interface  loopback0                              // 设置 loop 口
RouteA_config_l0#ip address 10.10.10.1 255.255.255.0
RouteA_config_l0#ip ospf network point-to-point                 // 还原 loop 口地址
```

路由器 B：

```
RouteB#config
RouteB_config#interface loopback0
RouteB_config_l0#ip address  10.10.11.1 255.255.255.0
RouteB_config_l0#ip ospf network point-to-point
```

步骤 2：验证接口配置，代码如下。

```
RouteA #sh interface loopback0
Loopback0 is up, line protocol is up
  Hardware is Loopback
  Interface address is 10.10.10.1/24
  MTU 1514 bytes, BW 8000000 kbit, DLY 500 usec
  Encapsulation LOOPBACK
RouteB #sh interface loopback0
Loopback0 is up, line protocol is up
  Hardware is Loopback
  Interface address is 10.10.11.1/24
  MTU 1514 bytes, BW 8000000 kbit, DLY 500 usec
  Encapsulation LOOPBACK
```

步骤 3：路由器的 OSPF 配置，代码如下。

A 的配置：

RouteA_config#*router ospf 1*　　　　　　// 启动 OSPF 进程，进程号为 1

RouteA_config_ospf_1#*network 10.10.10.0 255.255.255.0 area 0*// 注意要写掩码和区域号

RouteA_config_ospf_1#*network 192.168.1.0 255.255.255.0 area 0*

B 的配置：

RouteB_config#*router ospf 1*

RouteB_config_ospf_1#*network 10.10.11.0 255.255.255.0 area 0*

RouteB_config_ospf_1#*network 192.168.1.0 255.255.255.0 area 0*

步骤 4：查看路由表，代码如下。

路由器 A：

RouteA#*sh ip route*

Codes: C - connected, S - static, R - RIP, B - BGP, BC - BGP connected

　　　　D - DEIGRP, DEX - external DEIGRP, O - OSPF, OIA - OSPF inter area

　　　　ON1 - OSPF NSSA external type 1, ON2 - OSPF NSSA external type 2

　　　　OE1 - OSPF external type 1, OE2 - OSPF external type 2

　　　　DHCP - DHCP type

VRF ID: 0

C　　　10.10.10.0/24　　　　　　is directly connected, Loopback0

O　　　10.10.11.1/32　　　　　　[110,1600] via 192.168.1.2(on Serial0/2)

　　　　　　　　　　　　　　　　　　　　// 注意到环回接口产生的位置是主机路由

C　　　192.168.1.0/24　　　　　is directly connected, Serial0/2

路由器 B：

RouteB#*show ip route*

Codes: C - connected, S - static, R - RIP, B - BGP, BC - BGP connected

　　　　D - DEIGRP, DEX - external DEIGRP, O - OSPF, OIA - OSPF inter area

　　　　ON1 - OSPF NSSA external type 1, ON2 - OSPF NSSA external type 2

　　　　OE1 - OSPF external type 1, OE2 - OSPF external type 2

　　　　DHCP - DHCP type

VRF ID: 0

O　　　10.10.10.1/32　　　　　　[110,1601] via 192.168.1.1(on Serial0/1)// 注意管理距离为 110

C　　　10.10.11.0/24　　　　　　is directly connected, Loopback0

C　　　192.168.1.0/24　　　　　is directly connected, Serial0/1

步骤 5：其他验证命令，代码如下。

RouteB#*show ip ospf 1*　　　　　　　// 显示该 OSPF 进程的信息

OSPF process: 1, Router ID: 10.10.11.1

Distance: intra-area 110, inter-area 110, external 150

SPF schedule delay 5 secs, Hold time between two SPFs 10 secs

SPFTV:0(0), TOs:6, SCHDs:6
All Rtrs support Demand-Circuit.
Number of areas is 1
AREA: 0
　Number of interface in this area is 2(UP: 1)
　Area authentication type: None
　All Rtrs in this area support Demand-Circuit.

RouteA#***show ip ospf interface***　　　// 显示 OSPF 接口状态和类型
Loopback0 is up, line protocol is up
　Internet Address: 10.10.10.1/24
　Interface index: 9
　Nettype: Point-to-Point　　// 环回接口的网络类型为点对点
　OSPF process is 1,　AREA: 0, Router ID: 10.10.10.1
　Cost: 1, Transmit Delay is 1 sec, Priority 0
　Hello interval is 10, Dead timer is 40, Retransmit is 5
　OSPF INTF State is ILOOPBACK
　Neighbor Count is 0, Adjacent neighbor count is 0

Serial0/2 is up, line protocol is up
　Internet Address: 192.168.1.1/24
　Interface index: 3
　Nettype: Point-to-Point
　OSPF process is 1,　AREA: 0, Router ID: 10.10.10.1
　Cost: 1600, Transmit Delay is 1 sec, Priority 1
　Hello interval is 10, Dead timer is 40, Retransmit is 5
　OSPF INTF State is IPOINT_TO_POINT
　Neighbor Count is 1, Adjacent neighbor count is 1
　　Adjacent with neighbor 192.168.1.2

RouteA#***sh ip ospf neighbor***　　　// 显示 OSPF 邻居

　　　　　　　　　　　　　　OSPF process: 1
　　　　　　　　　　　　　　　AREA: 0

Neighbor ID	Pri	State	DeadTime	Neighbor Addr	Interface
10.10.11.1	1	FULL/-	37	192.168.1.2	Serial0/2

步骤 6：查看路由器 B 的路由表，代码如下。
RouteB#***show ip route***

```
Codes: C - connected, S - static, R - RIP, B - BGP, BC - BGP connected
       D - BEIGRP, DEX - external BEIGRP, O - OSPF, OIA - OSPF inter area
       ON1 - OSPF NSSA external type 1, ON2 - OSPF NSSA external type 2
       OE1 - OSPF external type 1, OE2 - OSPF external type 2
       DHCP - DHCP type, L1 - IS-IS level-1, L2 - IS-IS level-2
VRF ID: 0
O     10.10.10.1/32         [110,1601] via 192.168.1.1(on Serial0/1)
C     10.10.11.0/24         is directly connected, Loopback0
C     192.168.1.0/24        is directly connected, Serial0/1
```

注意事项和排错

1）每个路由器的 OSPF 进程号可以不同，一个路由器可以有多个 OSPF 进程。
2）OSPF 是无类路由协议，一定要加掩码。
3）第一个区域必须是区域 0。

相关命令详解

相关命令详解见表 6-7。

表 6-7

命 令	解 释
ip ospf network point-to-point	还原 loop 口掩码
router ospf 1	启动 OSPF 进程，进程号为 1
network 10.10.10.0 255.255.255.0 area 0	IP 地址、子网掩码和区域号

共同思考

1）OSPF 与 RIP 有哪些区别？
2）环回接口有什么好处？

第 7 章 路由器接口设置实践

实训 1 路由器串口 PPP-PAP 配置

实训目标

今天小刘的领导为小刘分配了一个任务,为一个局端用户分配广域网链路接口,并针对广域网用户配置 PAP 方式的验证。在运营商网络中,一般使用串口实现广域网链路的互联。小刘先在公司使用两台路由器采用串口背靠背互联的方式,模拟了运营商用户间的广域网链路 PAP 认证的配置。

1) 掌握 PAP 验证配置。
2) 理解验证过程。

通过学习本实训,了解路由器串口 PPP-PAP,具备配置广域网链路 PAP 认证的能力。

实训设备

1) DCR2655/2659/2855 路由器 2 台。
2) CR-V35FC 1/ CR-V35MT 1 各 1 条。

实训拓扑

实训拓扑图如图 7-1 所示。

图 7-1

实训任务

步骤 1:基本配置如下。

RouteA		RouteB	
接口	IP 地址	接口	IP 地址
S0/2 DCE	192.168.1.1	S0/1 DTE	192.168.1.2
账号	密码	账号	密码
RouteA	digitalchinaA	RouteB	digitalchinaB

步骤 2：路由器 A 的配置，代码如下。

```
Router>enable                                          // 进入特权模式
Router #config                                         // 进入全局配置模式
Router_config#hostname RouteA                          // 修改机器名
RouteA_config#aaa authentication ppp test local
RouteA_config#username RouteB password digitalchinaB   // 设置账号密码
RouteA_config#interface s0/2                           // 进入接口模式
RouteA_config_s0/2#ip address 192.168.1.1 255.255.255.0   // 配置 IP 地址
RouteA_config_s0/2#encapsulation PPP                   // 封装 PPP
RouteA_config_s0/2#ppp authentication pap test         // 设置验证方式
RouteA_config_s0/2#ppp pap sent-username RouteA password digitalchinaA
                                                       // 设置发送给对方验证的账号密码
Router-A_config_s0/2#physical-layer speed 64000        // 配置 DCE 时钟频率
Router-A_config_s0/2#no shutdown
Router-A_config_s0/2#^Z
```

步骤 3：查看路由器 A 的配置，代码如下。

```
RouteA#show interface s0/2                             // 查看接口状态
Serial0/2 is up, line protocol is down                 // 对端没有设置，所以协议是 DOWN
  Mode=Sync DCE Speed=64000                            // 查看 DCE
    DTR=UP, DSR=UP, RTS=UP, CTS=UP, DCD=UP
  MTU 1500 bytes, BW 64 kbit, DLY 2000 usec
  Interface address is 192.168.1.1/24                  // 查看 IP 地址
  Encapsulation PPP, loopback not set                  // 查看封装协议
  Keepalive set(10 sec)
  LCP  Starting configuration exchange
  PAP  Closed,  Message: 'none'
  IPCP Listening -- waiting for remote host to attempt open
       local IP address: 192.168.1.1  remote IP address: 0.0.0.0
  60 second input rate 0 bits/sec, 0 packets/sec!
  60 second output rate 0 bits/sec, 0 packets/sec!
    52 packets input, 1248 bytes, 4 unused_rx, 0 no buffer
    0 input errors, 0 CRC, 0 frame, 0 overrun, 0 ignored, 0 abort
    76 packets output, 1800 bytes, 8 unused_tx, 0 underruns
error:
    0 clock, 0 grace
PowerQUICC SCC specific errors:
    0 recv allocb mblk fail      0 recv no buffer
    0 transmitter queue full     0 transmitter hwqueue_full
```

步骤 4：路由器 B 的配置，代码如下。

```
Router>enable                                          // 进入特权模式
Router#config                                          // 进入全局配置模式
Router_config#hostname RouteB                          // 修改机器名
RouteB_config#aaa authentication ppp test local
RouteB_config#username RouteA password digitalchinaA   // 设置账号密码
RouteB_config#interface s0/1                           // 进入接口模式
RouteB_config_s0/1#ip address 192.168.1.2 255.255.255.0   // 配置 IP 地址
RouteB_config_s0/1#encapsulation PPP                   // 封装 PPP
RouteB_config_s0/1#ppp authentication pap test         // 设置验证方式
RouteB_config_s0/1#ppp pap sent-username RouteB password digitalchinaB
                                                       // 设置发送给对方验证的账号密码
RouteB_config_s0/1#shutdown
RouteB_config_s0/1#no shutdown
RouteB_config_s0/1#^Z                                  // 按 <Ctrl + Z> 组合键进入特权模式
```

步骤 5：查看路由器 B 的配置，代码如下。

```
RouteB#show interface s0/1                             // 查看接口状态
Serial0/1 is up, line protocol is up                   // 接口和协议都是 up
 Mode=Sync DTE                                         // 查看 DTE
   DTR=UP, DSR=UP, RTS=UP, CTS=UP, DCD=UP
  MTU 1500 bytes, BW 64 kbit, DLY 2000 usec
  Interface address is 192.168.1.2/24                  // 查看 IP 地址
  Encapsulation PPP, loopback not set                  // 查看封装协议
  Keepalive set(10 sec)
  LCP  Opened
  PAP  Opened,  Message: 'Welcome to Digital China Networks Limited Router'
  IPCP Opened
        local IP address: 192.168.1.2  remote IP address: 192.168.1.1
 60 second input rate 26 bits/sec, 0 packets/sec!
 60 second output rate 23 bits/sec, 0 packets/sec!
    15 packets input, 394 bytes, 5 unused_rx, 0 no buffer
    0 input errors, 0 CRC, 0 frame, 0 overrun, 0 ignored, 0 abort
    14 packets output, 370 bytes, 8 unused_tx, 0 underruns
error:
    0 clock, 0 grace
PowerQUICC SCC specific errors:
    0 recv allocb mblk fail       0 recv no buffer
    0 transmitter queue full      0 transmitter hwqueue_full
```

步骤 6：检测连通性，代码如下。

RouteA#*ping 192.168.1.2*

PING 192.168.1.2 (192.168.1.2): 56 data bytes

!!!!!

--- 192.168.1.2 ping statistics ---

5 packets transmitted, 5 packets received, 0% packet loss

round-trip min/avg/max = 20/20/20 ms

RouteB#*ping 192.168.1.1*

PING 192.168.1.1 (192.168.1.1): 56 data bytes

!!!!!

--- 192.168.1.1 ping statistics ---

5 packets transmitted, 5 packets received, 0% packet loss

round-trip min/avg/max = 20/20/20 ms

注意事项和排错

1）账号密码一定要交叉对应，发送的账号密码要和对方账号数据库中的账号密码对应。

2）不要忘记配置 DCE 的时钟频率。

相关命令详解

ppp authentication

使用接口配置命令 ppp authentication 指定接口上使用 CHAP 或 PAP 协议的次序，使用 no ppp authentication 取消认证。

ppp authentication {**chap**|**ms-chap**|**pap**}[[*list-name*|**default**][**callin**]

no ppp authentication

相关命令详解见表 7-1。

表 7-1

命　　令	解　　释
chap	在串行接口上激活 CHAP
pap	在串行接口上激活 PAP
ms-chap	在串行接口上激活 MS-CHAP
list-name	（可选的）与 AAA/TACACS+ 一起使用，指定执行认证时使用的 TACACS ＋方法列表名。如果没有指定列表名，系统将使用默认列表。使用命令 aaa authentication ppp 创建列表
default	（可选的）与 AAA/TACACS+ 一起使用。使用命令 aaa authentication ppp 创建默认列表
callin	（可选的）指定仅对收到的呼叫（calls）进行认证

共同思考

1）在什么环境下需要配置验证方式？

2）PAP 验证是否非常安全？

实训 2　路由器串口 PPP-CHAP 配置

实训目标

小刘在模拟运营商到用户广域网接入的实现中发现，使用 PPP-PAP 方式虽然可以通过用户名和密码限制合法用户的接入，但 PAP 的交互过程采用的是明文认证，这样的认证方式很容易被破解和攻击。小刘在和用户交流后，建议用户采用 PPP-CHAP 的方式对广域网链路进行加密认证。

1）掌握 CHAP 验证配置。
2）理解验证过程。

通过学习本实训，了解路由器串口 PPP-CHAP，具备配置广域网链路 CHAP 认证的能力。

实训设备

1）DCR2655/2659/2855 路由器 2 台。
2）CR-V35FC 1/ CR-V35MT 1 各 1 条。

实训拓扑

实训拓扑图如图 7-2 所示。

图　7-2

实训任务

步骤 1：基本配置如下。

RouteA		RouteB	
接口	IP 地址	接口	IP 地址
S0/2 DCE	192.168.1.1	S0/1 DTE	192.168.1.2
账号	密码	账号	密码
RouteA	digitalchina	RouteB	digitalchina

步骤 2：路由器 A 的配置，代码如下。

Router>*enable*　　　　　　　　　　　　　　　　　　　　//进入特权模式

Router #*config*	// 进入全局配置模式
Router _config#*hostname RouteA*	// 修改机器名
RouteA_config#*aaa authentication ppp test local*	
	// 定义一个名为 test,使用本地数据进行验证的 aaa 验证方法
RouteA_config#*username RouteB password digitalchina*	// 设置账号密码
RouteA_config#*interface s0/2*	// 进入接口模式
RouteA_config_s0/2#*ip address 192.168.1.1 255.255.255.0*	// 配置 IP 地址
RouteA_config_s0/2#*encapsulation PPP*	// 封装 PPP
RouteA_config_s0/2#*ppp authentication chap test*	// 设置验证方式
RouteA_config_s0/2#*ppp chap hostname RouteA*	// 设置发送给对方验证的账号密码
Router-A_config_s0/2#*physical-layer speed 64000*	// 配置 DCE 时钟频率
Router-A_config_s0/2#*no shutdown*	
Router-A_config_s0/2#*^Z*	

步骤 3:查看路由器 A 的配置,代码如下。

RouteA#*show interface s0/2*	// 查看接口状态
Serial0/2 is up, line protocol is down	// 由于对端没有设置,所以协议是 DOWN
Mode=Sync DCE Speed=64000	// 查看 DCE
DTR=UP, DSR=UP, RTS=UP, CTS=UP, DCD=UP	
MTU 1500 bytes, BW 64 kbit, DLY 2000 usec	
Interface address is 192.168.1.1/24	// 查看 IP 地址
Encapsulation PPP, loopback not set	// 查看封装协议
Keepalive set(10 sec)	
LCP Starting configuration exchange	
PAP Closed, Message: 'none'	
IPCP Listening -- waiting for remote host to attempt open	
local IP address: 192.168.1.1 remote IP address: 0.0.0.0	
60 second input rate 0 bits/sec, 0 packets/sec!	
60 second output rate 0 bits/sec, 0 packets/sec!	
52 packets input, 1248 bytes, 4 unused_rx, 0 no buffer	
0 input errors, 0 CRC, 0 frame, 0 overrun, 0 ignored, 0 abort	
76 packets output, 1800 bytes, 8 unused_tx, 0 underruns	
error:	
0 clock, 0 grace	
PowerQUICC SCC specific errors:	
0 recv allocb mblk fail 0 recv no buffer	
0 transmitter queue full 0 transmitter hwqueue_full	

步骤 4:路由器 B 的配置,代码如下。

Router>*enable*	// 进入特权模式
Router #*config*	// 进入全局配置模式

```
Router_config#hostname RouteB                              // 修改机器名
RouteB_config#aaa authentication ppp test local
RouteB_config#username RouteA password digitalchina        // 设置账号密码
RouteB_config#interface s0/1                               // 进入接口模式
RouteB_config_s0/1#ip address 192.168.1.2 255.255.255.0    // 配置 IP 地址
RouteB_config_s0/1#encapsulation PPP                       // 封装 PPP
RouteB_config_s0/1#ppp authentication chap test            // 设置验证方式
RouteB_config_s0/1#ppp chap hostname RouteB                // 设置发送给对方验证的账号密码
RouteB_config_s0/1#shutdown
RouteB_config_s0/1#no shutdown
RouteB_config_s0/1#^Z                                      // 按 <Ctrl + Z> 组合键进入特权模式
```

步骤 5：查看路由器 B 的配置，代码如下。

```
RouteB#show interface s0/1                                 // 查看接口状态
Serial0/1 is up, line protocol is up                       // 接口和协议都是 up
  Mode=Sync DTE                                            // 查看 DTE
  DTR=UP, DSR=UP, RTS=UP, CTS=UP, DCD=UP
  MTU 1500 bytes, BW 64 kbit, DLY 2000 usec
  Interface address is 192.168.1.2/24                      // 查看 IP 地址
  Encapsulation PPP, loopback not set                      // 查看封装协议
  Keepalive set(10 sec)
  LCP Opened
  PAP Opened, Message: 'Welcome to Digital China Networks Limited Router'
  IPCP Opened
        local IP address: 192.168.1.2  remote IP address: 192.168.1.1
  60 second input rate 26 bits/sec, 0 packets/sec!
  60 second output rate 23 bits/sec, 0 packets/sec!
    15 packets input, 394 bytes, 5 unused_rx, 0 no buffer
    0 input errors, 0 CRC, 0 frame, 0 overrun, 0 ignored, 0 abort
    14 packets output, 370 bytes, 8 unused_tx, 0 underruns
error:
    0 clock, 0 grace
PowerQUICC SCC specific errors:
    0 recv allocb mblk fail          0 recv no buffer
    0 transmitter queue full         0 transmitter hwqueue_full
```

步骤 6：检测连通性，代码如下。

```
RouteA#ping 192.168.1.2
PING 192.168.1.2 (192.168.1.2): 56 data bytes
!!!!!
```

```
--- 192.168.1.2 ping statistics ---
5 packets transmitted, 5 packets received, 0% packet loss
round-trip min/avg/max = 20/20/20 ms
RouteB#ping 192.168.1.1
PING 192.168.1.1 (192.168.1.1): 56 data bytes
!!!!!
--- 192.168.1.1 ping statistics ---
5 packets transmitted, 5 packets received, 0% packet loss
round-trip min/avg/max = 20/20/20 ms
```

注意事项和排错

1）双方密码一定要一致，发送的账号要和对方账号数据库中的账号对应。
2）不要忘记配置 DCE 的时钟频率。

相关命令详解

相关命令详解见表 7-2。

表 7-2

命 令	解 释
ppp chap hostname RouteB	设置发送给对方验证的账号
encapsulation PPP	封装 PPP
ppp authentication chap test	设置验证方式

共同思考

1）CHAP 和 PAP 这两种验证有什么不同？
2）CHAP 验证是否非常安全？

第 8 章 路由器应用技术实践

实训 1 标准访问列表配置

实训目标

一天小张接到某用户打来的电话,用户说内网中有一台 PC,由于 PC 上运行了重要的财务数据,用户不希望内网中其他网段的用户访问到这台 PC。小张想到用户的这种需求可以使用访问列表来控制用户间数据互访而实现。为了更好地解决用户的问题,小张在向用户答复前利用手边的路由器进行了标准访问列表的实验。

禁止 192.168.0.0/24 对 PC2 的访问。

通过学习本实训,能够掌握标准访问列表的概念,具备配置 ACL 功能的能力。

实训设备

1)DCR2655/2659/2855 路由器 2 台。
2)CR-V35FC 1/ CR-V35MT 1 各 1 条,交叉双绞线 2 条。
3)PC 2 台。

实训拓扑

实训拓扑图如图 8-1 所示。

图 8-1

实训任务

步骤 1:参照表 8-1,配置所有接口的地址,并测试连通性。

表 8-1

主 机 名	F0/0	S0/1(DTE)	S0/2(DCE)	PC 地址	默认网关
RouteA	192.168.0.1/24	—	192.168.1.1/24		
RouteB	192.168.2.1/24	192.168.1.2/24	—		
PC1	—	—	—	192.168.0.2/24	192.168.0.1
PC2	—	—	—	192.168.2.2/24	192.168.2.1

```
RouteA#ping 192.168.1.2
PING 192.168.1.2 (192.168.1.2): 56 data bytes
!!!!!
--- 192.168.1.2 ping statistics ---
5 packets transmitted, 5 packets received, 0% packet loss
round-trip min/avg/max = 20/28/30 ms
```

步骤 2：配置静态路由，代码如下。

```
RouteA#show ip route
Codes: C - connected, S - static, R - RIP, B - BGP, BC - BGP connected
       D - BEIGRP, DEX - external BEIGRP, O - OSPF, OIA - OSPF inter area
       ON1 - OSPF NSSA external type 1, ON2 - OSPF NSSA external type 2
       OE1 - OSPF external type 1, OE2 - OSPF external type 2
       DHCP - DHCP type, L1 - IS-IS level-1, L2 - IS-IS level-2
VRF ID: 0
C       192.168.0.0/24          is directly connected, FastEthernet0/0
C       192.168.1.0/24          is directly connected, Serial0/2
S       192.168.2.0/24          [1,0] via 192.168.1.2(on Serial0/2)


RouteB#show ip route
Codes: C - connected, S - static, R - RIP, B - BGP, BC - BGP connected
       D - BEIGRP, DEX - external BEIGRP, O - OSPF, OIA - OSPF inter area
       ON1 - OSPF NSSA external type 1, ON2 - OSPF NSSA external type 2
       OE1 - OSPF external type 1, OE2 - OSPF external type 2
       DHCP - DHCP type, L1 - IS-IS level-1, L2 - IS-IS level-2
VRF ID: 0
S       192.168.0.0/24          [1,0] via 192.168.1.1(on Serial0/1)
C       192.168.1.0/24          is directly connected, Serial0/1
C       192.168.2.0/24          is directly connected, FastEthernet0/0
```

步骤 3：PC1 能与 PC2 通信，如图 8-2 所示。

图 8-2

步骤 4：配置访问控制列表禁止 PC1 所在的网段对 PC2 的访问，代码如下。

```
RouteB#config
RouteB_config#ip access-list standard 1                         //定义标准的访问控制列表
RouteB_config_std_nacl#deny 192.168.0.0 255.255.255.0           //基于源地址
```

RouteB_config_std_nacl#*permit any*   // 因为有隐含的 deny any

步骤 5：将访问控制列表（ACL）绑定在相应的接口，代码如下。

RouteB_config#*int f0/0*   // 进入离目标最近的接口
RouteB_config_f0/0#*ip access-group 1 out*   // 绑定 ACL 1 在出接口的方向

步骤 6：验证，代码如下。

RouteB#*sh ip access-list*
Standard IP access list 1
   deny   192.168.0.0 255.255.255.0
   permit any

步骤 7：测试结果，如图 8-3 所示。

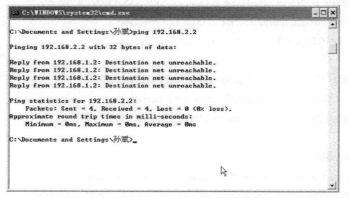

图 8-3

注意事项和排错

1）标准访问控制列表是基于源地址的。
2）每条访问控制列表都有隐含的拒绝。
3）标准访问控制列表一般绑定在离目标最近的接口。
4）注意方向，以该接口为参考点，IN 是流进的方向；OUT 是流出的方向。

相关命令详解

deny 命令参数见表 8-2。

表 8-2

参数	参数说明
protocol	协议名或 IP 号。它可以是关键字 icmp、igmp、igrp、ip、ospf、tcp 或 udp，也可以是表 IP 号的 0～255 的一个整数。为了匹配任何 Internet 协议（包括 ICMP、TCP 和 UDP）使用关键字 IP。某些协议允许进一步限定
source	源网络或主机号。有两种方法指定源：32 位二进制数，用四个点隔开的十进制数表示。使用关键字 any 作为 0.0.0.0 0.0.0.0 的源和源掩码缩写
source-mask	源地址网络掩码。使用关键字 any 作为 0.0.0.0 0.0.0.0 的源和源掩码缩写
destination	目标网络或主机号。有两种方法指定： ①使用四个点隔开的十进制数表示的 32 位二进制数 ②使用关键字 any 作为 0.0.0.0 0.0.0.0 的目标和目标掩码的缩写
destination-mask	目标地址网络掩码。使用关键字 any 作为 0.0.0.0 0.0.0.0 的目标地址和目标地址掩码缩写

(续)

参　数	参　数　说　明
precedence	（可选）数据包可以由优先级过滤，用 0～7 的数字指定
tos	（可选）数据包可以使用服务层过滤。使用数字 0～15 指定
icmp-type	（可选）ICMP 包可由 ICMP 报文类型过滤。类型是数字 0～255
igmp-type	（可选）IGMP 包可由 IGMP 报文类型或报文名过滤。类型是 0～15 的数字
operator	（可选）比较源或目标端口。操作包括 lt（小于），gt（大于），eq（等于），neq（不等于）。如果操作符放在 source 和 source-mask 之后，那么它必须匹配这个源端口。如果操作符放在 destination 和 destination-mask 之后，那么它必须匹配目标端口
port	（可选）TCP 或 UDP 端口的十进制数字或名称。端口号是一个 0～65 535 的数字。TCP 端口名列在"使用方针"部分。当过滤 TCP 时，可以只使用 TCP 端口名称。UDP 端口名称也列在"使用说明"部分。当过滤 TCP 时，只可使用 TCP 端口名。当过滤 UDP 时，只可使用 UDP 端口名
established	（可选）只对 TCP，表示一个已建立的连接。当 TCP 数据报 ACK 或 RST 位已经设置时，表示一个已建立的连接。ACK 或 RST 位非设置时即非匹配的情况是进行初始化 TCP 数据报，以重新形成一个连接
log	（可选）可以进行日志记录

　　可以使用访问列表控制包在接口上的传输，控制虚拟终端线路访问以及限制路由选择更新的内容。在匹配发生以后停止检查扩展的访问表。分段 IP 包，而不是初始段，立即由任何扩展的 IP 访问列表接收。扩展的访问列表用于控制访问虚拟终端线路或限制路由选择更新的内容，不必匹配 TCP 源端口、服务值的类型或包的优先权。

ip access-group

参数见表 8-3。

表　8-3

参　数	参　数　说　明
access-list-name	访问表名。这是一个最长为 20 个字符的字符串
in	在进接口时使用访问列表
out	在出接口时使用访问列表

　　访问列表既可用在出接口，也可用在入接口。对于标准的入口访问列表，在接收到包之后，对照访问列表检查包的源地址。对于扩展的访问列表，该路由器也检查目标地址。如果访问表允许该地址，那么软件继续处理该包。如果访问表不允许该地址，则该软件放弃包并返回一个 ICMP 主机不可到达报文。

　　对于标准的出口访问表，在接收和路由一个包到控制接口以后，软件对照访问列表检查包的源地址。对于扩展的访问列表，路由器还要检查接收端地址。如果访问列表允许该地址，软件则传送这个包。如果访问列表不允许该地址，则软件放弃这个包并返回一个 ICMP 主机不可达报文。

　　如果指定的访问列表不存在，则所有的包允许通过。

ip access-list

参数见表 8-4。

表　8-4

参　数	参　数　说　明
standard	指定为标准访问列表
extended	指定为扩展访问列表
name	访问列表名。这是一个最长 20 的字符串

使用此命令将进入 IP 访问列表配置模式，在 IP 访问列表配置模式中，可以用 deny 或 permit 命令来配置访问规则。

permit

参数见表 8-5。

表 8-5

参 数	参 数 说 明
protocol	协议名或 IP 号。它可以是关键字 icmp、igmp、igrp、ip、ospf、tcp 或 udp，也可以是表 IP 协议号的 0～255 的一个整数。为了匹配任何 Internet 协议（包括 ICMP、TCP 和 UDP）使用关键字 IP。某些协议允许进一步限定
source	源网络或主机号。有两种方法指定源：32 位二进制数，用四个点隔开的十进制数表示。使用关键字 any 作为 0.0.0.0 0.0.0.0 的源和源掩码缩写
source-mask	源地址网络掩码。使用关键字 any 作为 0.0.0.0 0.0.0.0 的源和源掩码缩写
destination	目标网络或主机号。有两种方法指定： ①使用四个点隔开的十进制数表示的 32 位二进制数 ②使用关键字 any 作为 0.0.0.0 0.0.0.0 的目标和目标掩码的缩写
destination-mask	目标地址网络掩码。使用关键字 any 作为 0.0.0.0 0.0.0.0 的目标地址和目标地址掩码缩写
precedence	（可选）包可以由优先级过滤，用 0～7 的数字指定
tos	（可选）数据包可以使用服务层过滤。使用数字 0～15 指定
icmp-type	（可选）ICMP 包可由 ICMP 报文类型过滤。类型是 0～255 的数字
igmp-type	（可选）IGMP 包可由 IGMP 报文类型或报文名过滤。类型是 0～15 的数字
operator	（可选）比较源或目标端口。操作包括 lt（小于），gt（大于），eq（等于），neq（不等于）。如果操作符放在 source 和 source-mask 之后，那么它必须匹配这个源端口。如果操作符放在 destination 和 destination-mask 之后，那么它必须匹配目标端口
port	（可选）TCP 或 UDP 端口的十进制数字或名称。端口号是一个 0～65 535 的数字。TCP 端口名列在"使用方针"部分。当过滤 TCP 时，可以只使用 TCP 端口名称。UDP 端口名称也列在"使用说明"部分。当过滤 TCP 时，只可使用 TCP 端口名。当过滤 UDP 时，只可使用 UDP 端口名
established	（可选）只对 TCP，表示一个已建立的连接。当 TCP 数据报 ACK 或 RST 位已经设置时，表示一个已建立的连接。ACK 或 RST 位非设置时即非匹配的情况是进行初始化 TCP 数据报，以重新形成一个连接
log	（可选）可以进行日志记录

可以使用访问列表控制包在接口上的传输，控制虚拟终端线路访问以及限制路由选择更新的内容。在匹配发生以后停止检查扩展的访问表。

分段 IP 包，而不是初始段，立即由任何扩展的 IP 访问表接收。扩展的访问表用于控制访问虚拟终端线路或限制路由选择更新的内容，不必匹配 TCP 源端口、服务值的类型或包的优先权。

共同思考

1）为什么访问控制列表最后要加一条允许？

2）除了绑定在 F0/0 以外，在现在的环境中还能绑定在哪个接口上？什么方向？

实训 2　扩展访问列表配置

实训目标

小刘在向用户回访网络运行状态的过程中，用户提到了一个新的需求。现在网络中有一台 PC2，用户不希望内网用户可以使用 Telnet 方式登录到 PC2。同时，内网中还有一台网管服务器 PC1，网管服务器 PC1 使用 ping 的方式能检验 PC2 是否在线。

小刘总结了用户的需求，针对 PC2 来说，一方面禁止内网用户的 Telnet，另一方面需要允许 PC1 的 ping 请求，这种场景因为涉及具体的应用，就需要使用扩展访问控制列表限制具体的应用端口了。

禁止 192.168.0.0/24 Telnet 到 PC2，但 PC1 能 ping 通 PC2。

通过学习本实训，能够掌握扩展访问列表的概念，具备配置扩展 ACL 功能的能力。

实训设备

1）DCR2655/2659/2855 路由器 2 台。
2）CR-V35FC 1/ CR-V35MT 1 各 1 条，交叉双绞线 2 条。
3）PC 2 台。

实训拓扑

实训拓扑图如图 8-4 所示。

图　8-4

实训任务

步骤 1：参照表 8-6，配置所有接口的地址，并测试连通性。

表　8-6

主机名	F0/0	S0/1（DTE）	S0/2（DCE）	PC 地址	默认网关
RouteA	192.168.0.1/24	—	192.168.1.1/24	—	—
RouteB	192.168.2.1/24	192.168.1.2/24	—	—	—
PC1	—	—	—	192.168.0.2/24	192.168.0.1
PC2	—	—	—	192.168.2.2/24	192.168.2.1

Router-A#*ping 192.168.1.2*

PING 192.168.1.2 (192.168.1.2): 56 data bytes

!!!!!

--- 192.168.1.2 ping statistics ---

5 packets transmitted, 5 packets received, 0% packet loss

round-trip min/avg/max = 20/28/30 ms

步骤 2：配置静态路由，代码如下。

RouteA#*show ip route*

Codes: C - connected, S - static, R - RIP, B - BGP, BC - BGP connected
       D - BEIGRP, DEX - external BEIGRP, O - OSPF, OIA - OSPF inter area
       ON1 - OSPF NSSA external type 1, ON2 - OSPF NSSA external type 2
       OE1 - OSPF external type 1, OE2 - OSPF external type 2
       DHCP - DHCP type, L1 - IS-IS level-1, L2 - IS-IS level-2

VRF ID: 0

C   192.168.0.0/24   is directly connected, FastEthernet0/0
C   192.168.1.0/24   is directly connected, Serial0/2
S   192.168.2.0/24   [1,0] via 192.168.1.2(on Serial0/2)

RouteB#*show ip route*

Codes: C - connected, S - static, R - RIP, B - BGP, BC - BGP connected
       D - BEIGRP, DEX - external BEIGRP, O - OSPF, OIA - OSPF inter area
       ON1 - OSPF NSSA external type 1, ON2 - OSPF NSSA external type 2
       OE1 - OSPF external type 1, OE2 - OSPF external type 2
       DHCP - DHCP type, L1 - IS-IS level-1, L2 - IS-IS level-2

VRF ID: 0

S   192.168.0.0/24   [1,0] via 192.168.1.1(on Serial0/1)
C   192.168.1.0/24   is directly connected, Serial0/1
C   192.168.2.0/24   is directly connected, FastEthernet0/0

步骤 3：PC1 能与 PC2 通信，如图 8-5 所示。

图 8-5

步骤 4：在路由器 A 上设置访问列表，代码如下。

RouteA#*conf*

RouteA_config#*ip access-list extended 192*   // 定义扩展访问列表

RouteA_config_ext_nacl#*deny tcp 192.168.0.0 255.255.255.0 192.168.2.2 255.255.255.0 eq 23*

                                                           // 设置扩展访问列表，拒绝 Telnet

RouteA_config_ext_nacl#*permit icmp any any*   // 允许 ping

RouteA_config_ext_nacl#*exit*

```
RouteA_config#int f0/0                                  // 进入离源比较近的接口
RouteA_config_f0/0#ip access-group 192 in               // 绑定访问列表在 IN 的方向
```

步骤 5：查看访问列表，代码如下。

```
Router-A#sh ip access-list
Extended IP access list 192
    deny    tcp 192.168.0.0 255.255.255.0 192.168.2.2 255.255.255.0 eq telnet
    permit icmp any any
```

步骤 6：验证，代码如下。

```
C:\Documents and Settings\Administrator>ping 192.168.2.2

Pinging 192.168.2.2 with 32 bytes of data:

Reply from 192.168.2.2: bytes=32 time=22ms TTL=62
Reply from 192.168.2.2: bytes=32 time=21ms TTL=62
Reply from 192.168.2.2: bytes=32 time=21ms TTL=62
Reply from 192.168.2.2: bytes=32 time=21ms TTL=62

Ping statistics for 192.168.2.2:
    Packets: Sent = 4, Received = 4, Lost = 0 (0% loss),
Approximate round trip times in milli-seconds:
    Minimum = 21ms, Maximum = 22ms, Average = 21ms

C:\Documents and Settings\Administrator>telnet 192.168.2.2
正在连接到 192.168.2.2... 不能打开到主机的连接，在端口 23: 连接失败
```

注意事项和排错

1）扩展访问列表通常放在离源比较近的地方。
2）扩展访问列表可以基于源、目标 IP、协议、端口号等条件过滤。
3）命令比较长，可以通过？的方式查看帮助。
4）注意隐含的 deny。

共同思考

1）为什么扩展访问列表通常放在离源比较近的接口处？
2）访问列表中主机还有什么表示方法？
3）为什么访问列表中至少要有一条允许规则？

实训 3　路由器 NAT 实训

实训目标

NAT 网络地址转换

小宋明天要去现场为用户架设一台出口路由器，想到出口路由器作为网络的出口设备，要为内网用户提供 NAT（网络地址转换）功能，就在本地先使用路由器进行 NAT 功能的配置实验。

通过学习本实训，能够掌握路由器 NAT 的概念，具备配置 NAT 功能的能力。

实训设备

1) DCR2655/2659/2855 路由器 2 台。
2) CR-V35FC 1/ CR-V35MT 1 各 1 条，交叉双绞线 2 条。
3) PC 2 台。

实训拓扑

实训拓扑图如图 8-6 所示。

图 8-6

实训任务

步骤 1：按表 8-7 将接口地址和 PC 地址配置好，并且做连通性测试。

表 8-7

设 备	S0/1	F0/0	IP 地址	网 关
Router A	192.168.1.1/24	192.168.0.1/24	—	—
Router B	192.168.1.2/24	192.168.2.1/24	—	—
PC 1	—	—	192.168.0.3/24	192.168.0.1
SERVER	—	—	192.168.2.2/24	192.168.2.1

内部的 PC 需要访问外部的服务器。

假设在路由器 A 上做地址转换，将 192.168.0.0/24 转换成 192.168.1.10 ～ 192.168.1.20 之间的地址，并且做端口的地址复用。

步骤 2：配置路由器 A 的 NAT，代码如下。

```
Router A#conf
Router A_config#ip access-list standard 1                 // 定义访问控制列表
Router A_config_std_nacl#permit 192.168.0.0 255.255.255.0        // 定义允许转换的源地址范围
Router A_config_std_nacl#exit
Router A_config#ip nat pool overld 192.168.1.10 192.168.1.20 255.255.255.0    // 定义名为 overld 的转换地址池
Router A_config#ip nat inside source list 1 pool overld overload    // 配置将 ACL 允许的源地址转换成 overld 中的地址，并且做端口的地址复用
Router A_config#int f0/0
Router A_config_f0/0#ip nat inside                 // 定义 F0/0 为内部接口
Router A_config_f0/0#int s0/1
Router A_config_s1/1#ip nat outside                // 定义 S0/1 为外部接口
Router A_config_s1/1#exit
Router A_config#ip route 0.0.0.0 0.0.0.0 192.168.1.2   // 配置路由器 A 的默认路由
```

步骤 3：查看路由器 B 的路由表，代码如下。

```
Router B#sh ip route
Codes: C - connected, S - static, R - RIP, B - BGP, BC - BGP connected
       D - DEIGRP, DEX - external DEIGRP, O - OSPF, OIA - OSPF inter area
       ON1 - OSPF NSSA external type 1, ON2 - OSPF NSSA external type 2
       OE1 - OSPF external type 1, OE2 - OSPF external type 2
       DHCP - DHCP type
VRF ID: 0
// 注意：并没有到 192.168.0.0 的路由
```

步骤 4：测试连通性，如图 8-7 所示。

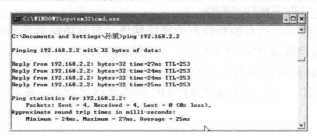

图 8-7

步骤 5：查看地址转换表，代码如下。

```
Router A#sh ip nat translations
Pro. Dir    Inside local        Inside global       Outside local       Outside global

ICMP OUT 192.168.0.3:512    192.168.1.10:12512   192.168.1.2:12512   192.168.1.2:12512
C        192.168.1.0/24      is directly connected, Serial0/1
C        192.168.2.0/24      is directly connected, FastEthernet0/0
```

注意事项和排错

1）注意转换的方向和接口。
2）注意地址池、ACL 的名称。
3）需要配置 A 的默认路由。

相关命令详解

ip nat inside source

参数见表 8-8。

表 8-8

参数	参数说明
access-list-name	IP 访问列表的名字。源地址符合访问列表的报文将被用地址池中的全局地址来翻译
pool name	地址池的名字，从这个池中动态地分配全局 IP 地址
interface type number	指定网络接口
overload	（可选）使路由器对多个本地地址使用一个全局的地址。当 overload 被设置后，相同或者不同主机的多个会话将被用 TCP 或 UDP 端口号来区分

(续)

参　数	参　数　说　明
static local-ip	建立一条独立的静态地址翻译；参数为分配给内部网主机的本地 IP 地址。这个地址可以自由选择，或从 RFC 1918 中分配
local-port	设置本地 TCP/UDP 端口号，范围为 1～65 535
static global-ip	建立一条独立的静态地址翻译；这个参数为内部主机建立一个外部的网络可以唯一访问的 IP 地址
global-port	设置全局 TCP/UDP 端口号，范围为 1～65 535
tcp	设置 TCP 端口翻译
udp	设置 UDP 端口翻译
network local-network	设置本地网段翻译
global-network	设置全局网段翻译
mask	设置网段翻译的网络掩码

这个命令有两种形式：动态的和静态的地址翻译。带有访问列表的格式建立动态翻译。来自和标准访问列表相匹配的地址的报文，将用指定的池中分配的全局地址来进行地址翻译，这个池是用 ip nat pool 命令所指定的。

可以作为替代方法，为带有关键字 STATIC 的语法格式创建一条独立的静态地址翻译。

共同思考

1）为什么在 B 上不需要配置 192.168.0.0 的路由就能够通信？
2）为什么在 A 上需要配置默认路由？
3）如果外部接口地址是通过拨号动态获得的，那么该如何配置？

实训 4　IPSec VPN（IKE）的配置

VPN 工作原理及配置

实训目标

用户由于业务开展，需要在另一个城市设立办事处，为了使总部和办事处使用同一个平台办公。用户向运营商提出了在总部和办事处间建立 VPN 连接的请求。小李在接到用户的请求后，与用户沟通确定总部和办事处间使用公网链路互联，传输数据时有数据加密的需求。

基于上面的需求，小李认为使用 IPSec VPN 的方式可以更好地实现用户的需求。小李使用本地的两台路由器模拟了用户的应用环境，开始了 IPSec VPN 的配置。

在路由器 A 与 B 之间建立 VPN，保护从 PC1 到 SERVER 的数据。

通过学习本实训，能够掌握 IPSec VPN 的概念，具备配置 IPSec VPN 功能的能力。

实训设备

1）DCR2655/2659/2855 路由器 2 台。
2）CR-V35FC 1/ CR-V35MT 1 各 1 条，交叉双绞线 2 条。
3）PC 2 台。

实训拓扑

实训拓扑图如图 8-8 所示。

图 8-8

实训任务

步骤 1：参照表 8-9，配置所有接口的地址，并测试连通性。

表 8-9

设 备	F0/0	S0/1（DTE）	S0/2（DCE）	PC 地址	默 认 网 关
RouteA	192.168.0.1/24	—	192.168.1.1/24	—	—
RouteB	192.168.2.1/24	192.168.1.2/24	—	—	—
PC1	—	—	—	192.168.0.2/24	192.168.0.1
SERVER	—	—	—	192.168.2.2/24	192.168.2.1

步骤 2：路由器 A 的配置，代码如下。

RouteA#*conf*
RouteA_config#*ip access-list extended 101*   //确定要经过 VPN 保护的数据流
RouteA_config_ext_nacl#*permit ip 192.168.0.0 255.255.255.0 192.168.2.0 255.255.255.0*
RouteA_config_ext_nacl#*exit*
RouteA_config#*ip route 0.0.0.0 0.0.0.0 192.168.1.2*   //配置静态路由
RouteA_config#*crypt isakmp policy 10*   //配置 IKE 策略
RouteA_config_isakmp#*authentication pre-share*   //设置认证方式
RouteA_config_isakmp#*encryption des*   //设置加密方式
RouteA_config_isakmp#*hash md5*   //设置数字签名算法
RouteA_config_isakmp#*group 1*   //设置 DH 方式
RouteA_config_isakmp#*lifetime 86400*   //设置生存期
RouteA_config_isakmp#*exit*
RouteA_config#*crypto isakmp key digital 192.168.1.2*   //设置共用密钥
RouteA_config#*crypto ipsec transform-set one*   //设置变换集
RouteA_config_crypto_trans#*transform-type esp-des esp-md5-hmac*   //ESP 加密和验证
RouteA_config_crypto_trans#*mode tunnel*   //设置为隧道模式
RouteA_config_crypto_trans#*exit*
RouteA_config#*crypto map my 10 ipsec-isakmp*   //配置 IPSec 加密映射
RouteA_config_crypto_map#*set transform-set one*   //关联变换集
RouteA_config_crypto_map#*set peer 192.168.1.2*   //设置对等体地址

```
RouteA_config_crypto_map#match address 101          // 关联需要加密的数据流
RouteA_config_crypto_map#exit
RouteA_config#interface s0/2                        // 进入 VPN 的接口
RouteA_config_s0/2#crypto map my                    // 绑定 IPSec 加密映射
RouteA_config_s0/2#^Z
```

步骤 3：查看配置，代码如下。

```
RouteA#sh crypto isakmp policy                      // 查看 IKE 策略
Protection suite of priority 10
        encryption algorithm:    DES  - Data Encryption Standard (56 bit keys).
        hash algorithm:          Message Digest 5
        authentication method:   Pre-Shared Key
        Diffie-Hellman group:    #1 (768 bit)
        lifetime:                86400 seconds
Default protection suite
        encryption algorithm:    DES  - Data Encryption Standard (56 bit keys).
        hash algorithm:          Secure Hash Standard
        authentication method:   Pre-Shared Key
        Diffie-Hellman group:    #1 (768 bit)
        lifetime:                86400 seconds


RouteA#sh crypto isakmp sa                 // 查看 IKE 安全关联（没有建立，为空）


RouteA#sh crypto map                       // 查看 IPSec 映射
Crypto Map my 10 ipsec-isakmp
    Extended IP access list 101
      permit ip 192.168.0.0 255.255.255.0 192.168.2.0 255.255.255.0
    peer = 192.168.1.2
    PFS (Y/N): N
    Security association lifetime: 4608000 kilobytes/3600 seconds
    Transform sets={ one,}


Router-A#sh crypto ipsec sa                          // 查看 IPSec 关联
Interface: Serial0/2
Crypto map name:my , local addr. 192.168.1.1
 local  ident (addr/mask/prot/port): (192.168.0.0/255.255.255.0/0/0)
 remote ident (addr/mask/prot/port): (192.168.2.0/255.255.255.0/0/0)
 local crypto endpt.: 192.168.1.1,  remote crypto endpt.: 192.168.1.2


Router-A#sh crypto ipsec transform-set              // 查看变换集
Transform set one: { esp-des esp-md5-hmac }
   will negotiate ={ Tunnel }
```

步骤 4：路由器 B 的配置，代码如下。

```
RouteB>ena
RouteB#conf
RouteB_config#ip access-list extended 101
RouteB_config_ext_nacl#permit ip 192.168.2.0 255.255.255.0 192.168.0.0 255.255.255.0
RouteB_config_ext_nacl#exit
RouteB_config#ip route 192.168.0.0 255.255.255.0 192.168.1.1
RouteB_config#crypto isakmp policy 10                          //注意与 A 要一致
RouteB_config_isakmp#authentication pre-share
RouteB_config_isakmp#hash md5
RouteB_config_isakmp#encryption des
RouteB_config_isakmp#group 1
RouteB_config_isakmp#lifetime 86400
RouteB_config_isakmp#exit
RouteB_config#crypto isakmp key digital 192.168.1.1            //注意与 A 要一致
RouteB_config#crypto ipsec transform-set one
RouteB_config_crypto_trans#transform-type esp-des esp-md5-hmac //注意与 A 要一致
RouteB_config_crypto_trans#mode tunnel
RouteB_config_crypto_trans#exit
RouteB_config#crypto map my 10 ipsec-isakmp                    //注意与 A 要一致
RouteB_config_crypto_map#set transform-set one
RouteB_config_crypto_map#set peer 192.168.1.1
RouteB_config_crypto_map#match address 101
RouteB_config_crypto_map#exit
RouteB_config#int s1/0
RouteB_config_s1/0#crypto map my
RouteB_config_s1/0#^Z
```

步骤 5：查看配置，代码如下。

```
Router-B#sh crypto isakmp policy
Protection suite of priority 10
        encryption algorithm:    DES    - Data Encryption Standard (56 bit keys).
        hash algorithm:          Message Digest 5
        authentication method:   Pre-Shared Key
        Diffie-Hellman group:    #1 (768 bit)
        lifetime:                86400 seconds
Default protection suite
        encryption algorithm:    DES    - Data Encryption Standard (56 bit keys).
        hash algorithm:          Secure Hash Standard
        authentication method:   Pre-Shared Key
        Diffie-Hellman group:    #1 (768 bit)
        lifetime:                86400 secondsRouter-B#sh crypto isakmp sa
```

```
Router-B#sh crypto ipsec sa
Interface: Serial0/1
Crypto map name:my ,    local addr. 192.168.1.2
local   ident (addr/mask/prot/port): (192.168.2.0/255.255.255.0/0/0)
  remote ident (addr/mask/prot/port): (192.168.0.0/255.255.255.0/0/0)
  local crypto endpt.: 192.168.1.2,  remote crypto endpt.: 192.168.1.1


Router-B#sh crypto ipsec transform-set
Transform set one: { esp-des esp-md5-hmac }
      will negotiate ={ Tunnel }
Router-B#sh crypto map
Crypto Map my 10 ipsec-isakmp
      Extended IP access list 101
        permit ip 192.168.2.0 255.255.255.0 192.168.0.0 255.255.255.0
      peer = 192.168.1.1
      PFS (Y/N): N
      Security association lifetime: 4608000 kilobytes/3600 seconds
      Transform sets={ one,}
```

步骤 6：测试结果，如图 8-9 所示。

图　8-9

再次查看安全关联，代码如下。

```
Router-B#sh crypto isakmp sa
       dst              src            state          state-id       conn
   192.168.1.1      192.168.1.2     <R>Q_SA_SETUP        2           2 my 10
   192.168.1.1      192.168.1.2     <R>M_SA_SETUP        1           2 my 10


Router-B#sh crypto ipsec sa
Interface: Serial0/1
Crypto map name:my ,    local addr. 192.168.1.2
local  ident (addr/mask/prot/port): (192.168.2.0/255.255.255.0/0/0)
```

```
    remote ident (addr/mask/prot/port): (192.168.0.0/255.255.255.0/0/0)
    local crypto endpt.: 192.168.1.2,    remote crypto endpt.: 192.168.1.1
    inbound esp sas:
      spi:0x7f73737b(2138272635)
        transform:  esp-des  esp-md5-hmac
        in use settings ={ Tunnel }
        sa timing: remaining key lifetime (k/sec):    (4607998/3495)
    outbound esp sas:
      spi:0x2cb62d75(750136693)
        transform:  esp-des  esp-md5-hmac
        in use settings ={ Tunnel }
        sa timing: remaining key lifetime (k/sec):    (4607998/3495)
```

注意事项和排错

1) 注意两端参数要一致。
2) ACL 的作用是确定哪些数据需要经过 VPN。

相关命令详解

transform-type

加密变换配置状态下,要设置变换类型,使用 transform-type 命令。

transform-type transform1 [transform2[transform3]]

参数说明见表 8-10。

表 8-10

参　数	参 数 说 明
transform1/ transform2/ transform3	可以指定 3 个以下的变换。这些变换定义了 IPSec 安全协议和算法。可接受的变换值在"使用说明"中详细阐述

默认:默认的变换类型为 ESP-DES(ESP 采用 DES 加密算法)。
命令模式:加密变换配置状态。
使用说明:

变换集合可以指定一个或两个 IPSec 安全协议(或 ESP,或 AH,或两者都有),并且可以指定已选定的安全协议使用哪种算法。

变换集合的定义可以指定 1 ~ 3 个变换——每个变换代表一个 IPSec 安全协议(ESP 或 AH)和想要使用的算法的组合。当 IPSec 安全联盟协商时使用了某一个变换集合,整个变换集合(协议、算法和其他设置的组合)必须和对端的一个变换集合相匹配。

在一个变换集合中,可以指定 AH、ESP 或两者都指定。如果在变换集合中指定了一个 ESP,那么可以只定义 ESP 加密变换,也可以 ESP 加密变换和 ESP 验证变换两者都定义。

表 8-11 中显示了可行的变换组合。

IPSec 协议：ESP 和 AH。

ESP 和 AH 协议都为 IPSec 提供了安全服务。

ESP 提供了分组加密，以及可选的数据验证和抗重播服务。

AH 提供了数据验证和抗重播服务。

ESP 使用一个 ESP 头和一个 ESP 尾对受保护数据——或是一个完整的 IP 自寻址数据包（或仅是有效负载）——进行封装。AH 是嵌入在受保护数据中的；它将一个 AH 头直接插入在外部 IP 头后、内部 IP 数据包或有效负载前。隧道模式中要对整个 IP 数据报文进行封装和保护，而传送模式中只对 IP 数据报文中的有效负载进行封装和保护。

表 8-11

AH 变换中选择一种		ESP 加密变换中选择一种		ESP 验证变换中选择一种	
变换	描述	变换	描述	变换	描述
ah-md5-hmac	带 MD5(HMAC 变量) 的 AH 验证算法	esp-des	采用 DES 的 ESP 加密算法	esp-md5-hmac	带 MD5(HMAC 变量) 的 ESP 验证算法
ah-sha-hmac	带 SHA(HMAC 变量) 的 AH 验证算法	esp-3des	采用 3DES 的 ESP 加密算法	esp-sha-hmac	带 SHA(HMAC 变量) 的 ESP 验证算法

1）选择适当的变换。

IPSec 变换比较复杂。下面的提示能够帮助用户选择适合自己情况的变换：

如果想要提供数据机密性，那么可以使用 ESP 加密变换。

如果想要提供对外部 IP 报头以及数据的验证，那么可以使用 AH 变换。

如果使用一个 ESP 加密变换，那么可以考虑使用 ESP 验证变换或 AH 变换来提供变换集合的验证服务。

如果想要数据验证功能（或使用 ESP 或使用 AH），可以选择 MD5 或 SHA 验证算法。SHA 算法比 MD5 要更安全，但速度更慢。

2）加密变换配置状态。

在执行了 crypto ipsec transform-set 命令以后，就将进入加密变换配置状态。在这种状态下，可以将模式改变到隧道模式或传输模式。在做完这些改变以后，输入 exit 来返回到全局配置状态下。

3）改变现存的变换。

如果在 transform-type 命令中为一个变换集合指定一个或多个变换，那么指定的这些变换将会替换掉变换集合中现存的变换。如果改变了 transform-type，则改变将只被运用到引用了此变换集合的加密映射表上。但改变将不会被运用到现存的安全联盟上，而会被用于新建立的安全联盟。如果要使新的设置立即生效，则可以使用 clear crypto sa 命令来清除安全联盟数据库的部分或全部数据。

示例：

以下例子定义了一个变换集合。

```
crypto ipsec transform-set one
transform-type esp-des esp-sha-hmac
```

实训 5 L2TP/PPTP VPN 的配置

VPN 工作原理及配置

实训目标

小宋的表哥在一家网吧负责网管工作。最近的一次聚会中，表哥提到网管的工作主要是维护网吧内网络设备及 PC 的正常运行，但经常出现他已经回到家中，网吧反馈有问题的情况。一般遇到这种情况，表哥就要赶回单位进行网络排查和调试。表哥想向小宋咨询一下有没有什么方法可以更加简便地针对远程网络进行维护。

小宋：表哥，这种情况对于网络维护人员或公司内有移动办公的用户非常常见，一般通过 L2TP/PPTP VPN 就可以远程登录到内网进行维护管理了。因为 L2TP/PPTP 是 Windows 自带的标准 VPN 客户端协议，只要是 Windows 操作系统的 PC 都可以支持这两种 VPN 协议。

于是小宋就向表哥演示了 L2TP/PPTP VPN 的配置过程。

通过学习本实训，能够掌握 L2TP/PPTP VPN 的概念，具备配置 L2TP/PPTP VPN 功能的能力。

实训设备

1）DCR2655/2659/2855 路由器 2 台。
2）CR-V35FC 1/ CR-V35MT 1 各 1 条，交叉双绞线 2 条。
3）PC 2 台。

实训拓扑

实训拓扑图如图 8-10 所示。

图 8-10

实训任务

步骤 1：参照表 8-12，配置所有接口的地址，并测试连通性。

表 8-12

设备	F0/0	S0/1（DTE）	S0/2（DCE）	PC 地址	默认网关
RouteA	192.168.0.1/24	—	192.168.1.1/24	—	—
RouteB	192.168.2.1/24	192.168.1.2/24	—	—	—
PC1	—	—	—	192.168.0.2/24	192.168.0.1
PC2	—	—	—	192.168.2.2/24	192.168.2.1

步骤 2：路由器 A 的配置，代码如下。

RouteA#*conf*

RouteA_config#*int virtual-tunnel 0*

RouteA_config_vn0#*ip address 172.16.1.2 255.255.255.0*

RouteA_config_vn0#*ppp chap host test@dcn.net*

RouteA_config_vn0#*ppp chap password 1234*

RouteA_config_vn0#*exit*

RouteA_config#vpdn *enable*

RouteA_config#*vpdn-group 0*

RouteA_config_vpdn#*request-dialin*

RouteA_config_vpdn#*protocol l2tp*

RouteA_config_vpdn#*initiate-to ip 192.168.1.2 priority 1*

RouteA_config_vpdn#*domain dcn.net*

RouteA_config_vn0#*exit*

RouteA_config#*ip route 192.168.2.0 255.255.255.0 virtual-tunnel 0*

步骤 3：查看路由器 A 的配置，代码如下。

RouteA#*sh l2tp tunnel*

L2TP Tunnel Information:

No active tunnels

RouteA#*sh l2tp session*

L2TP Session Information:

No active sessions

RouteA#*sh int virtual-tunnel 0*

Virtual-tunnel0 is up, line protocol is down

   Hardware is Unknown device

   MTU 1500 bytes, BW 100000 kbit, DLY 10000 usec

   Interface address is 172.16.1.2/24

   Encapsulation PPP, loopback not set

   Keepalive set(10 sec)

   LCP   Starting configuration exchange

   IPCP Listening -- waiting for remote host to attempt open

      local IP address: 172.16.1.2 remote IP address: 0.0.0.0

步骤4：路由器B的配置，代码如下。

```
RouteB#conf
RouteB_config#user test@dcn.net password 0 1234
RouteB_config#ip local pool l2tppool 172.16.1.10 10
RouteB_config#int virtual-template 0
RouteB_config_vt0#ip address 172.16.1.1 255.255.255.0
RouteB_config_vt0#ppp authen chap
RouteB_config_vt0#peer default ip address pool l2tppool
RouteB_config_vt0#exit
RouteB_config#vpdn enable
RouteB_config#vpdn-group 0
RouteB_config_vpdn#accept-dialin
RouteB_config_vpdn#protocol l2tp
RouteB_config_vpdn#lcp-renegotiation
RouteB_config_vpdn#port virtual-template 0
RouteB_config_vpdn#exit
RouteB_config#ip route 192.168.0.0 255.255.255.0 172.16.1.2
RouteB_config#^Z
```

步骤5：查看路由器B的配置。

```
RouteB# sh run
Building configuration...

Current configuration:
!
!version 1.3.2E
service timestamps log date
service timestamps debug date
no service password-encryption
!
hostname Router-B
!
ip host a 192.168.1.1
ip host c 192.168.2.2
!
!
!
!
ip local pool l2tppool 172.16.1.10 10
!
aaa authentication ppp test local
```

```
username test@dcn.net password 0 1234
!
!
!
interface Virtual-template0
  ip address 172.16.1.1 255.255.255.0
  no ip directed-broadcast
  ppp authentication chap test
  peer default ip address pool l2tppool
!
interface FastEthernet0/0
  ip address 192.168.2.1 255.255.255.0
  no ip directed-broadcast
!
interface Serial1/0
  ip address 192.168.1.2 255.255.255.0
  no ip directed-broadcast
!
interface Async0/0
  no ip address
  no ip directed-broadcast
!
!
!
!
ip route 192.168.0.0 255.255.255.0 172.16.1.2
!
!
!
!
!
!
!
!
vpdn enable
!
vpdn-group 0
  accept-dialin
```

lcp-renegotiation

protocol l2tp

local-name Digitalchina

virtual-template 0

步骤 6：测试结果，如图 8-11 所示。

```
C:\Documents and Settings\孙斌>ping 192.168.2.2 -t

Pinging 192.168.2.2 with 32 bytes of data:

Request timed out.
Request timed out.
Request timed out.
Request timed out.
Request timed out.
Reply from 192.168.2.2: bytes=32 time=26ms TTL=253
Reply from 192.168.2.2: bytes=32 time=23ms TTL=253
Reply from 192.168.2.2: bytes=32 time=23ms TTL=253
Reply from 192.168.2.2: bytes=32 time=23ms TTL=253
Reply from 192.168.2.2: bytes=32 time=24ms TTL=253
Reply from 192.168.2.2: bytes=32 time=24ms TTL=253
Reply from 192.168.2.2: bytes=32 time=24ms TTL=253
Reply from 192.168.2.2: bytes=32 time=23ms TTL=253
Reply from 192.168.2.2: bytes=32 time=23ms TTL=253
Reply from 192.168.2.2: bytes=32 time=23ms TTL=253
```

图 8-11

注意事项和排错

1）注意两端参数要一致。
2）ACL 的作用是确定哪些数据需要经过 VPN。
3）密钥要交叉对应。

第 9 章 路由器综合应用进阶

实训1 综合实验1

实训目标

根据网络拓扑:

1) 路由器背对背线缆两端 IP 地址分别为 1.1.1.1/24 和 1.1.1.2/24。PPP 封装,无验证。

2) 路由器 R1 与交换机 S1 连接使用 F0/0 端口,交换机 S1 的 E0/0/23 端口与路由器 R1F0/0 均为 Trunk 端口。

3) 交换机 S1 与 S2 之间的链路也为 Trunk 链路。

4) VLAN10 成员 IP 地址范围为 192.168.1.0/24,其中 R1 虚拟 VLAN10 接口地址为 192.168.1.1/24。VLAN20 成员 IP 地址范围为 192.168.2.0/24,R1 虚拟 VLAN20 接口地址为 192.168.2.1/24。

5) R2 使用 F0/0 与 PC5 连接,PC5 的地址范围为 192.168.3.30/24,路由器接口地址为 192.168.3.1/24。

6) 实训中保证各 VLAN 中的 PC 与 192.168.3.30 都可以直接连通。

7) 路由器之间使用 RIP 的直连引入方式进行设置。

实训设备

1) DCR2655/2659/2855 路由器 2 台。
2) CR-V35FC 1/ CR-V35MT 1 各 1 条,交叉双绞线 2 条,直通线 5 条。
3) S4600 交换机 2 台。
4) PC 5 台。

实训拓扑

实训拓扑图如图 9-1 所示。

图 9-1

实训任务

步骤1：设置两台交换机 VLAN 并对封装端口进行配置，代码如下。

S1：

```
switch#config
switch(Config)#hostname S1
S1(Config)#vlan 10
S1 (Config-Vlan10)#switchport interface ethernet 0/0/1-9
Set the port Ethernet0/0/1 access vlan 10 successfully
Set the port Ethernet0/0/2 access vlan 10 successfully
Set the port Ethernet0/0/3 access vlan 10 successfully
Set the port Ethernet0/0/4 access vlan 10 successfully
Set the port Ethernet0/0/5 access vlan 10 successfully
Set the port Ethernet0/0/6 access vlan 10 successfully
Set the port Ethernet0/0/7 access vlan 10 successfully
Set the port Ethernet0/0/8 access vlan 10 successfully
Set the port Ethernet0/0/9 access vlan 10 successfully
S1 (Config-Vlan10)#exit
S1 (Config)#vlan 20
S1 (Config-Vlan20)#switchport interface ethernet 0/0/10-20
Set the port Ethernet0/0/10 access vlan 20 successfully
Set the port Ethernet0/0/11 access vlan 20 successfully
Set the port Ethernet0/0/12 access vlan 20 successfully
Set the port Ethernet0/0/13 access vlan 20 successfully
Set the port Ethernet0/0/14 access vlan 20 successfully
Set the port Ethernet0/0/15 access vlan 20 successfully
Set the port Ethernet0/0/16 access vlan 20 successfully
Set the port Ethernet0/0/17 access vlan 20 successfully
Set the port Ethernet0/0/18 access vlan 20 successfully
Set the port Ethernet0/0/19 access vlan 20 successfully
Set the port Ethernet0/0/20 access vlan 20 successfully
```

```
S1 (Config-Vlan20)#exit
S1 (Config)#interface ethernet 0/0/23
S1 (Config-Ethernet0/0/23)#switchport mode trunk
Set the port Ethernet0/0/23 mode TRUNK successfully
S1 (Config-Ethernet0/0/23)#switchport trunk native vlan 10
Set the port Ethernet0/0/23 native vlan 10 successfully
S1 (Config-Ethernet0/0/23)#switchport trunk allowed vlan all
set the port Ethernet0/0/23 allowed vlan successfully
S1 (Config-Ethernet0/0/23)#exit
S1 (Config)#interface ethernet 0/0/24
S1 (Config-Ethernet0/0/24)#switchport mode trunk
Set the port Ethernet0/0/24 mode TRUNK successfully
S1 (Config-Ethernet0/0/24)#switchport trunk native vlan 10
Set the port Ethernet0/0/24 native vlan 10 successfully
S1 (Config-Ethernet0/0/24)#switchport trunk allowed vlan all
set the port Ethernet0/0/24 allowed vlan successfully
S1 (Config-Ethernet0/0/24)#
```

S2：

```
switch(Config)#hostname S2
S2(Config)#no int vlan 1
S2 (Config)#vlan 10
S2 (Config-Vlan10)#switchport interface ethernet 0/0/1-9
Set the port Ethernet0/0/1 access vlan 10 successfully
Set the port Ethernet0/0/2 access vlan 10 successfully
Set the port Ethernet0/0/3 access vlan 10 successfully
Set the port Ethernet0/0/4 access vlan 10 successfully
Set the port Ethernet0/0/5 access vlan 10 successfully
Set the port Ethernet0/0/6 access vlan 10 successfully
Set the port Ethernet0/0/7 access vlan 10 successfully
Set the port Ethernet0/0/8 access vlan 10 successfully
Set the port Ethernet0/0/9 access vlan 10 successfully
S2 (Config-Vlan10)#exit
S2 (Config)#vlan 20
S2 (Config-Vlan20)#switchport interface ethernet 0/0/10-20
Set the port Ethernet0/0/10 access vlan 20 successfully
Set the port Ethernet0/0/11 access vlan 20 successfully
Set the port Ethernet0/0/12 access vlan 20 successfully
Set the port Ethernet0/0/13 access vlan 20 successfully
Set the port Ethernet0/0/14 access vlan 20 successfully
Set the port Ethernet0/0/15 access vlan 20 successfully
Set the port Ethernet0/0/16 access vlan 20 successfully
Set the port Ethernet0/0/17 access vlan 20 successfully
Set the port Ethernet0/0/18 access vlan 20 successfully
```

```
Set the port Ethernet0/0/19 access vlan 20 successfully
Set the port Ethernet0/0/20 access vlan 20 successfully
S2 (Config-Vlan20)#exit
S2 (Config)#interface ethernet 0/0/24
S2 (Config-Ethernet0/0/24)#switchport mode trunk
Set the port Ethernet0/0/24 mode TRUNK successfully
S2 (Config-Ethernet0/0/24)#switchport trunk native vlan 10
Set the port Ethernet0/0/24 native vlan 10 successfully
S2 (Config-Ethernet0/0/24)#switchport trunk allowed vlan all
set the port Ethernet0/0/24 allowed vlan successfully
S2 (Config-Ethernet0/0/24)#exit
S2 (Config)#
```

验证配置。

S1：

```
S1#show vlan
```

VLAN	Name	Type	Media	Ports	
1	default	Static	ENET	Ethernet0/0/21	Ethernet0/0/22
				Ethernet0/0/23	Ethernet0/0/24(T)
				Ethernet0/0/25	Ethernet0/0/26
				Ethernet0/0/27	Ethernet0/0/28
10	VLAN0010	Static	ENET	Ethernet0/0/1	Ethernet0/0/2
				Ethernet0/0/3	Ethernet0/0/4
				Ethernet0/0/5	Ethernet0/0/6
				Ethernet0/0/7	Ethernet0/0/8
				Ethernet0/0/9	Ethernet0/0/23(T)
				Ethernet0/0/24	
20	VLAN0020	Static	ENET	Ethernet0/0/10	Ethernet0/0/11
				Ethernet0/0/12	Ethernet0/0/13
				Ethernet0/0/14	Ethernet0/0/15
				Ethernet0/0/16	Ethernet0/0/17
				Ethernet0/0/18	Ethernet0/0/19
				Ethernet0/0/20	Ethernet0/0/23(T)
				Ethernet0/0/24(T)	

S2：

```
S2#show vlan
```

VLAN	Name	Type	Media	Ports

1	default	Static	ENET	Ethernet0/0/21	Ethernet0/0/22
				Ethernet0/0/23	Ethernet0/0/24(T)
				Ethernet0/0/25	Ethernet0/0/26
				Ethernet0/0/27	Ethernet0/0/28
10	VLAN0010	Static	ENET	Ethernet0/0/1	Ethernet0/0/2
				Ethernet0/0/3	Ethernet0/0/4
				Ethernet0/0/5	Ethernet0/0/6
				Ethernet0/0/7	Ethernet0/0/8
				Ethernet0/0/9	Ethernet0/0/24
20	VLAN0020	Static	ENET	Ethernet0/0/10	Ethernet0/0/11
				Ethernet0/0/12	Ethernet0/0/13
				Ethernet0/0/14	Ethernet0/0/15
				Ethernet0/0/16	Ethernet0/0/17
				Ethernet0/0/18	Ethernet0/0/19
				Ethernet0/0/20	Ethernet0/0/24(T)

步骤 2：设置路由器以太网端口与 VLAN 成员可互通，代码如下。

```
Router#conf
Router_config#hostname R1
R1_config#interface f0/0.1     // 进入路由器子接口
R1_config_f0/0.1#ip address 192.168.1.1 255.255.255.0    // 为子接口添加 IP
R1_config_f0/0.1#encapsulation dot1Q 10    // 为 VLAN10 的数据包封装 802.1Q 协议
R1_config_f0/0.1#exit
R1_config#interface f0/0.2     // 进入路由器子接口
R1_config_f0/0.2#ip address 192.168.2.1 255.255.255.0    // 为子接口添加 IP
R1_config_f0/0.2#encapsulation dot1Q 20    // 为 VLAN20 的数据包封装 802.1Q 协议
R1_config_f0/0.2#exit
R1_config#
R1r#wr
Saving current configuration...
OK!
R1#
```

验证互通

分别将 PC 192.168.1.10 接入到左侧交换机的 E0/0/1 和右侧交换机的 E0/0/1，验证与 192.168.1.1 的连通性。

分别将 PC 192.168.2.10 接入到左侧交换机的 E0/0/10 和右侧交换机的 E0/0/10，验证与 192.168.2.1 的连通性。

步骤 3：配置路由器背对背链路，保证链路通畅。

左侧路由器：

```
R1#config
R1_config#interface serial 0/2
R1_config_s2/0#encapsulation ppp
```

```
R1_config_s2/0#physical-layer speed 64000
R1_config_s2/0#ip address 1.1.1.1 255.255.255.0
R1_config_s2/0#exit
R1_config#
```

右侧路由器:

```
Router#config
Router#hostname R2
R2_config#interface serial 0/1
R2_config_s1/0#encapsulation ppp
R2_config_s1/0#ip address 1.1.1.2 255.255.255.0
R2_config_s1/0#exit
R2_config#exit
R2#
```

验证连通性:

```
R2#ping 1.1.1.1
PING 1.1.1.1 (1.1.1.1): 56 data bytes
!!!!!
--- 1.1.1.1 ping statistics ---
5 packets transmitted, 5 packets received, 0% packet loss
round-trip min/avg/max = 20/22/30 ms
R2#
```

步骤4：配置右侧路由器以太网端口与 PC 192.168.3.30 可连通，代码如下。

```
R2#config
R2_config#interface fastethernet 0/0
R2_config_f0/0#ip address 192.168.3.1 255.255.255.0
R2_config_f0/0#exit
R2_config#exit
R2#2004-1-1 00:16:42 Configured from console 0 by DEFAULT
R2#
```

验证连通性:

```
R2#ping 192.168.3.30
PING 192.168.3.30 (192.168.3.30): 56 data bytes
!!!!!
--- 192.168.3.30 ping statistics ---
5 packets transmitted, 5 packets received, 0% packet loss
round-trip min/avg/max = 0/0/0 ms
R2#
```

步骤5：配置路由器的 RIP，验证连通结果。

左侧路由器（同右侧路由器）:

```
R1#config
```

```
R1_config#router rip
R1_config_rip#network 1.1.1.0
R1_config_rip#redistribute connect
R1_config_rip#exit
R1_config#exit
```

验证路由表。

左侧路由器：

```
R1#sh ip route
Codes: C - connected, S - static, R - RIP, B - BGP, BC - BGP connected
       D - DEIGRP, DEX - external DEIGRP, O - OSPF, OIA - OSPF inter area
       ON1 - OSPF NSSA external type 1, ON2 - OSPF NSSA external type 2
       OE1 - OSPF external type 1, OE2 - OSPF external type 2
       DHCP - DHCP type


VRF ID: 0
C        1.1.1.0/24           is directly connected, Serial0/2
R        1.1.1.1/32           [120,1] via 1.1.1.2(on Serial0/2)
C        1.1.1.2/32           is directly connected, Serial0/2
C        192.168.1.0/24       is directly connected, FastEthernet0/0.1
C        192.168.2.0/24       is directly connected, FastEthernet0/0.2
R1#
```

右侧路由器：

```
R2#sh ip route
Codes: C - connected, S - static, R - RIP, B - BGP, BC - BGP connected
       D - DEIGRP, DEX - external DEIGRP, O - OSPF, OIA - OSPF inter area
       ON1 - OSPF NSSA external type 1, ON2 - OSPF NSSA external type 2
       OE1 - OSPF external type 1, OE2 - OSPF external type 2
       DHCP - DHCP type


VRF ID: 0
C        1.1.1.0/24           is directly connected, Serial0/1
C        1.1.1.1/32           is directly connected, Serial0/1
R        1.1.1.2/32           [120,1] via 1.1.1.1(on Serial0/1)
R        192.168.1.0/24       [120,1] via 1.1.1.1(on Serial0/1)
R        192.168.2.0/24       [120,1] via 1.1.1.1(on Serial0/1)
C        192.168.3.0/24       is directly connected, FastEthernet0/0
R2#
```

验证 PC 的互通性（略）。

实训 2 综合实验 2

实训目标

根据网络拓扑：
1）PC1 与 PC2 能够互相 ping 通对方。
2）R1、R2、R3 之间使用 RIP 形成路由表。
3）PC1 与 PC2 之间的 FTP 数据使用 VPN 隧道通信，其他数据不必加密。
4）链路 R1-R2-R3 和 R1-R3 互为备份路由。

实训设备

1）DCR2655/2659/2855 路由器 3 台。
2）CR-V35FC 1/ CR-V35MT 1 各 1 条，交叉双绞线 4 条。
3）PC 2 台。

实训拓扑

实训拓扑图如图 9-2 所示。

图 9-2

实训任务

步骤 1：设置两台 PC 的地址。
步骤 2：配置路由器接口的地址。

R1：

Router#*conf*

Router#*hostname R1*

R1_config#*interface f0/0*

R1_config_f0/0#*ip address 192.168.1.1 255.255.255.0*

R1_config_f0/0#*no shutdown*

R1_config_f0/0#*exit*

R1_config#*interface g0/3*

R1_config_g0/3#*ip address 10.1.1.2 255.255.255.0*

R1_config_g0/3#*no shutdown*

R1_config_g0/3#*exit*

R1_config#*interface s0/2*

R1_config_s0/2#*ip address 10.1.3.1 255.255.255.0*

R1_config_s0/2#physical-layer speed 64000

R1_config_s0/2#*no shutdown*

R1_config_s0/2#*exit*

R1_config#

R2：

Router#*conf*

Router_config#*hostname R2*

R2_config#*interface g0/3*

R2_config_g0/3#*ip address 10.1.2.1 255.255.255.0*

R2_config_g0/3#*no shutdown*

R2_config_g0/3#*exit*

R2_config#*interface g0/4*

R2_config_g0/4#*ip address 10.1.1.1 255.255.255.0*

R2_config_g0/4#*no shutdown*

R2_config_g0/4#*exit*

R3：

Router#*conf*

Router_config#*hostname R3*

R3_config#*interface f0/0*

R3_config_f0/0#*ip address 192.168.10.1 255.255.255.0*

R3_config_f0/0#*no shutdown*

第 9 章 路由器综合应用进阶

```
R3_config_f0/0#exit
R3_config#interface g0/3
R3_config_g0/3#ip address 10.1.2.2 255.255.255.0
R3_config_g0/3#no shutdown
R3_config_g0/3#exit
R3_config#interface s0/1
R3_config_s0/1#ip address 10.1.3.2 255.255.255.0
R3_config_s0/1#no shutdown
R3_config_s0/1#exit
R3_config#
```

步骤 3：查看路由表。

R1：

```
R1#show ip route
Codes: C - connected, S - static, R - RIP, B - BGP, BC - BGP connected
       D - BEIGRP, DEX - external BEIGRP, O - OSPF, OIA - OSPF inter area
       ON1 - OSPF NSSA external type 1, ON2 - OSPF NSSA external type 2
       OE1 - OSPF external type 1, OE2 - OSPF external type 2
       DHCP - DHCP type, L1 - IS-IS level-1, L2 - IS-IS level-2
VRF ID: 0
C     10.1.1.0/24         is directly connected, GigaEthernet0/3
C     10.1.3.0/24         is directly connected, Serial0/2
C     192.168.1.0/24      is directly connected, FastEthernet0/0
```

R2：

```
R2#
R2#show ip route
Codes: C - connected, S - static, R - RIP, B - BGP, BC - BGP connected
       D - BEIGRP, DEX - external BEIGRP, O - OSPF, OIA - OSPF inter area
       ON1 - OSPF NSSA external type 1, ON2 - OSPF NSSA external type 2
       OE1 - OSPF external type 1, OE2 - OSPF external type 2
       DHCP - DHCP type, L1 - IS-IS level-1, L2 - IS-IS level-2
VRF ID: 0
C     10.1.1.0/24         is directly connected, GigaEthernet0/4
C     10.1.2.0/24         is directly connected, GigaEthernet0/3
```

R3：

```
R3#
R3#show ip route
Codes: C - connected, S - static, R - RIP, B - BGP, BC - BGP connected
       D - BEIGRP, DEX - external BEIGRP, O - OSPF, OIA - OSPF inter area
       ON1 - OSPF NSSA external type 1, ON2 - OSPF NSSA external type 2
       OE1 - OSPF external type 1, OE2 - OSPF external type 2
```

```
            DHCP - DHCP type, L1 - IS-IS level-1, L2 - IS-IS level-2
VRF ID: 0
C         10.1.2.0/24            is directly connected, GigaEthernet0/3
C         10.1.3.0/24            is directly connected, Serial0/1
C         192.168.10.0/24        is directly connected, FastEthernet0/0
R3#
```

步骤 4：配置路由器 RIP。

R1：

R1#*conf*

R1_config#*router rip*

R1_config_rip#*network 192.168.1.0*

R1_config_rip#*network 10.1.1.0*

R1_config_rip#*network 10.1.3.0*

R1_config_rip#*exit*

R1_config#

R2：

R2#*conf*

R2_config#*router rip*

R2_config_rip#*network 10.1.1.0*

R2_config_rip#*network 10.1.2.0*

R2_config_rip#*exit*

R2_config#

R3：

R3#conf

R3_config#*router rip*

R3_config_rip#*network 192.168.10.0*

R3_config_rip#*network 10.1.2.0*

R3_config_rip#*network 10.1.3.0*

R3_config_rip#*exit*

R3_config#

步骤 5：查看路由表。

R1：

R1#*show ip route*

Codes: C - connected, S - static, R - RIP, B - BGP, BC - BGP connected
       D - BEIGRP, DEX - external BEIGRP, O - OSPF, OIA - OSPF inter area
       ON1 - OSPF NSSA external type 1, ON2 - OSPF NSSA external type 2
       OE1 - OSPF external type 1, OE2 - OSPF external type 2
       DHCP - DHCP type, L1 - IS-IS level-1, L2 - IS-IS level-2

VRF ID: 0

C	10.1.1.0/24	is directly connected, GigaEthernet0/3
R	10.1.2.0/24	[120,1] via 10.1.3.2(on Serial0/2)
		[120,1] via 10.1.1.1(on GigaEthernet0/3)
C	10.1.3.0/24	is directly connected, Serial0/2
C	192.168.1.0/24	is directly connected, FastEthernet0/0
R	192.168.10.0/24	[120,1] via 10.1.3.2(on Serial0/2)

R2：

R2#*show ip route*

Codes: C - connected, S - static, R - RIP, B - BGP, BC - BGP connected
　　　D - BEIGRP, DEX - external BEIGRP, O - OSPF, OIA - OSPF inter area
　　　ON1 - OSPF NSSA external type 1, ON2 - OSPF NSSA external type 2
　　　OE1 - OSPF external type 1, OE2 - OSPF external type 2
　　　DHCP - DHCP type, L1 - IS-IS level-1, L2 - IS-IS level-2

VRF ID: 0

C	10.1.1.0/24	is directly connected, GigaEthernet0/4
C	10.1.2.0/24	is directly connected, GigaEthernet0/3
R	10.1.3.0/24	[120,1] via 10.1.2.2(on GigaEthernet0/3)
		[120,1] via 10.1.1.2(on GigaEthernet0/4)
R	192.168.1.0/24	[120,1] via 10.1.1.2(on GigaEthernet0/4)
R	192.168.10.0/24	[120,1] via 10.1.2.2(on GigaEthernet0/3)

R3：

R3#*show ip route*

Codes: C - connected, S - static, R - RIP, B - BGP, BC - BGP connected
　　　D - BEIGRP, DEX - external BEIGRP, O - OSPF, OIA - OSPF inter area
　　　ON1 - OSPF NSSA external type 1, ON2 - OSPF NSSA external type 2
　　　OE1 - OSPF external type 1, OE2 - OSPF external type 2
　　　DHCP - DHCP type, L1 - IS-IS level-1, L2 - IS-IS level-2

VRF ID: 0

R	10.1.1.0/24	[120,1] via 10.1.3.1(on Serial0/1)
		[120,1] via 10.1.2.1(on GigaEthernet0/3)
C	10.1.2.0/24	is directly connected, GigaEthernet0/3
C	10.1.3.0/24	is directly connected, Serial0/1
R	192.168.1.0/24	[120,1] via 10.1.3.1(on Serial0/1)
C	192.168.10.0/24	is directly connected, FastEthernet0/0

步骤 6：检验 PC 连通性，代码如下。

C:\Documents and Settings\Administrator>*ping 192.168.10.10*
Pinging 192.168.10.10 with 32 bytes of data:
Reply from 192.168.10.10: bytes=32 time=21ms TTL=62
Reply from 192.168.10.10: bytes=32 time=20ms TTL=62
Reply from 192.168.10.10: bytes=32 time=21ms TTL=62

```
Reply from 192.168.10.10: bytes=32 time=21ms TTL=62
Ping statistics for 192.168.10.10:
    Packets: Sent = 4, Received = 4, Lost = 0 (0% loss),
Approximate round trip times in milli-seconds:
    Minimum = 20ms, Maximum = 21ms, Average = 20ms
C:\Documents and Settings\Administrator>
```

步骤 7：配置并查看访问列表。

R1：

R1#*conf*

R1_config#*ip access-list extended for_vpn*

R1_config_ext_nacl#*permit tcp 192.168.1.10 255.255.255.255 192.168.10.10 255.255.255.255 eq 21*

R1_config_ext_nacl#*deny ip any any*

R1#*show ip access-lists*

Extended IP access list for_vpn

  permit tcp 192.168.1.10 255.255.255.255 192.168.10.10 255.255.255.255 eq ftp

  deny   ip any any

R1#

R3：

R3#*conf*

R3_config#*ip access-list extended for_vpn*

R3_config_ext_nacl#*permit tcp 192.168.10.10 255.255.255.255 192.168.1.10 255.255.255.255 eq 21*

R3_config_ext_nacl#*deny ip any any*

R3#*show ip access-lists*

Extended IP access list for_vpn

  permit tcp 192.168.10.10 255.255.255.255 192.168.1.10 255.255.255.255 eq ftp

  deny   ip any any

R3#

步骤 8：配置 VPN（通过串口）。

R1：

R1#*conf*

R1_config#

R1_config#*crypt isakmp key dcnu 10.1.3.2 255.255.255.0*

R1_config#*crypto ipsec transform-set r1*

R1_config_crypto_trans#*mode tunnel*

R1_config_crypto_trans#*transform-type esp-des esp-md5-hmac*

R1_config_crypto_trans#*exit*

```
R1_config#crypto map r1 10 ipsec-isakmp
R1_config_crypto_map#set transform-set r1
R1_config_crypto_map#set peer 10.1.3.2
R1_config_crypto_map#match address for_vpn
R1_config_crypto_map#exit
R1_config#interface s0/2
R1_config_s0/2#crypto map r1
R1_config_s0/2#^Z
```

R3:

```
R3#conf
R3_config#crypto isakmp key dcnu 10.1.3.1 255.255.255.0
R3_config#crypto ipsec transform-set r1
R3_config_crypto_trans#mode tunnel
R3_config_crypto_trans#transform-type esp-des esp-md5-hmac
R3_config_crypto_trans#exit
R3_config#crypto map r1 10 ipsec-isakmp
R3_config_crypto_map#set transform-set r1
R3_config_crypto_map#set peer 10.1.3.1
R3_config_crypto_map#match address for_vpn
R3_config_crypto_map#exit
R3_config#interface s0/1
R3_config_s0/1#crypto map r1
R3_config_s0/1#^Z
```

步骤9：配置 VPN 备份 (通过 G0/3 口)。

R1：

```
R1#conf
R1_config#crypt isakmp key dcnu 10.1.2.2 255.255.255.0
R1_config#crypto ipsec transform-set r2
R1_config_crypto_trans#transform-type esp-des esp-md5-hmac
R1_config_crypto_trans#exit
R1_config#crypto map r2 10 ipsec-isakmp
R1_config_crypto_map#set transform-set r2
R1_config_crypto_map#set peer 10.1.2.2
R1_config_crypto_map#match address for_vpn
R1_config_crypto_map#exit
R1_config#interface g0/3
R1_config_g0/3#crypto map r2
R1_config_g0/3#exit
```

R3:

```
R3#conf
R3_config#crypt isakmp key dcnu 10.1.1.2 255.255.255.0
R3_config#crypto ipsec transform-set r2
R3_config_crypto_trans#transform-type esp-des esp-md5-hmac
R3_config_crypto_trans#exit
R3_config#crypto map r2 10 ipsec-isakmp
R3_config_crypto_map#set transform-set r2
R3_config_crypto_map#set peer 10.1.1.2
R3_config_crypto_map#match address for_vpn
R3_config_crypto_map#exit
R3_config#interface g0/3
R3_config_g0/3#crypto map r2
R3_config_g0/3#exit
```

步骤10：验证结果。

R1#*show crypto ipsec sa*

Interface: Serial0/2

Crypto map name:r1, local addr. 10.1.3.1

local ident (addr/mask/prot/port): (192.168.1.10/255.255.255.255/6/0)

  remote ident (addr/mask/prot/port): (192.168.10.10/255.255.255.255/6/21)

  local crypto endpt.: 10.1.3.1, remote crypto endpt.: 10.1.3.2

Interface: GigaEthernet0/3

Crypto map name:r2, local addr. 10.1.1.2

local ident (addr/mask/prot/port): (192.168.1.10/255.255.255.255/6/0)

  remote ident (addr/mask/prot/port): (192.168.10.10/255.255.255.255/6/21)

  remote ident (addr/mask/prot/port): (192.168.10.10/255.255.255.255/6/21)

R3#*show crypto ipsec sa*

Interface: Serial0/1

Crypto map name:r1, local addr. 10.1.3.2

local ident (addr/mask/prot/port): (192.168.10.10/255.255.255.255/6/0)

  remote ident (addr/mask/prot/port): (192.168.1.10/255.255.255.255/6/21)

  local crypto endpt.: 10.1.3.2, remote crypto endpt.: 10.1.3.1

Interface: GigaEthernet0/3

Crypto map name:r2, local addr. 10.1.2.2

local ident (addr/mask/prot/port): (192.168.10.10/255.255.255.255/6/0)

  remote ident (addr/mask/prot/port): (192.168.1.10/255.255.255.255/6/21)

  local crypto endpt.: 10.1.2.2, remote crypto endpt.: 10.1.1.2

步骤 11：PC 互 ping。

C:\Documents and Settings\Administrator>*ping 192.168.10.10*
Pinging 192.168.10.10 with 32 bytes of data:
Reply from 192.168.10.10: bytes=32 time=12ms TTL=62
Reply from 192.168.10.10: bytes=32 time=12ms TTL=62
Reply from 192.168.10.10: bytes=32 time=11ms TTL=62
Reply from 192.168.10.10: bytes=32 time=11ms TTL=62

Ping statistics for 192.168.10.10:
　　Packets: Sent = 4, Received = 4, Lost = 0 (0% loss),
Approximate round trip times in milli-seconds:
　　Minimum = 11ms, Maximum = 12ms, Average = 11ms

此时通过串口实现，关掉串口，然后继续 ping，结果如下。

C:\Documents and Settings\Administrator>ping 192.168.10.10 -t
Pinging 192.168.10.10 with 32 bytes of data:
Reply from 192.168.1.1: Destination host unreachable.
Reply from 192.168.1.1: Destination host unreachable.
Reply from 192.168.1.1: Destination host unreachable.
Reply from 192.168.1.1: Destination host unreachable.
Reply from 192.168.1.1: Destination host unreachable.
Reply from 192.168.10.10: bytes=32 time=2ms TTL=61
Reply from 192.168.10.10: bytes=32 time=1ms TTL=61
Reply from 192.168.10.10: bytes=32 time=1ms TTL=61
Reply from 192.168.10.10: bytes=32 time=1ms TTL=61
Reply from 192.168.10.10: bytes=32 time=1ms TTL=61
Reply from 192.168.10.10: bytes=32 time=1ms TTL=61

Ping statistics for 192.168.10.10:
　　Packets: Sent = 22, Received = 22, Lost = 0 (0% loss),
Approximate round trip times in milli-seconds:
　　Minimum = 1ms, Maximum = 2ms, Average = 0ms

第3部分

无线实训

第 10 章 无线技术基础

实训 1 多 SSID 的无线接入配置

实训目标

无线技术基础

某公司部署无线网络以后，陆续因无线接入出现网络故障，影响了网络的稳定。经过分析研究，把公司办公无线接入和来宾接入隔离，使用两个 SSID 标识，负载均衡，提高网络安全。

配置要求：配置同一个 AP 广播两个 SSID，实现无线用户分别接入。

通过学习本实训，能够掌握无线网络概念，具备配置无线 SSID 功能的能力。

实训设备

1）DCWL-8200AP 1 台。
2）CS6200 三层交换机 1 台，双绞线 2 条。
3）PC 2 台。

实训拓扑

实训拓扑图如图 10-1 所示。

图 10-1

实训任务

步骤 1：在 AP 网关设备上配置（图 10-1 中的三层交换机）实现与 AP 的通信。

SW(config-if-ethernet0/0/1)#switchport mode trunk
//AP 上联口需要承载多个 VLAN 数据时需要将此链路模式更改为 Trunk 模式
SW(config-if-ethernet0/0/1)#switchport trunk native vlan 1
// 使 VLAN1 通过 Trunk 链路时不打 VLAN 标签，如果 AP 处于除 VLAN1，则其他 VLAN 就需要把

本征 VLAN 更改成该 VLAN

步骤 2：配置 VLAN。创建用户 VLAN、AP 的管理 VLAN，代码如下。

```
switch (config)#vlan 1                           //AP 的管理 VLAN，STA_A 也属于这个 VLAN
switch (config-vlan)#name Dcn_ap_ac              //VLAN 1 名称是 Dcn_ap_ac
switch (config-vlan)#vlan 10                     // 无线用户_B 所在 VLAN
switch (config-vlan)#name DCN_A                  //VLAN 10 名称是 DCN_A
```

步骤 3：配置 AP VLAN 和 STA VLAN 网关地址，代码如下。

```
switch (config)#interface vlan 1// 配置 AP 的网关
switch (config-if-vlan1)#ip address 192.168.1.254 255.255.255.0
switch (config)#interface vlan 10                // 配置 STA_A 的网关
switch (config-if-vlan1)#ip address 192.168.10.254 255.255.255.0
```

步骤 4：配置 AP 的 DHCP 服务器，代码如下。

```
switch (config)#service dhcp                     // 开启 DHCP 服务
switch (config)#ip dhcp pool Dcn_ap
// 创建 DHCP 地址池，名称是 Dcn_ap
switch (dhcp-wireless_ap-config)#network 192.168.1.0 255.255.255.0
// 分配给用户 STA_A 的地址网段
switch (dhcp-wireless_sta-config)#dns-server 114.114.114.114
// 分配给 STA_A 的 DNS 地址
switch (dhcp-wireless_ap-config)#default-router 192.168.1.254
// 分配给 STA_A 的网关地址
switch #show ip dhcp binding
// 查看 AP 获取的地址的相关信息
Total dhcp binding items: 1, the matched: 1
IP address          Hardware address         Lease expiration              Type
192.168.1.1         00-03-0F-19-99-80        Sun Jun 15 12:55:00 2014      Dynamic
```

步骤 5：配置 STA_B 的 DHCP 服务器，代码如下。

```
switch (config)#ip dhcp pool Dcn_sta
// 创建 DHCP 地址池，名称是 Dcn_sta
switch (dhcp-wireless_sta-config)#network 192.168.10.0 255.255.255.0
// 分配给 STA_B 的地址网段
switch (dhcp-wireless_sta-config)#dns-server 114.114.114.114
// 分配给 STA_B 的 DNS 地址
switch (dhcp-wireless_sta-config)#default-router 192.168.10.254
// 分配给 STA_B 的网关地址
```

步骤 6：配置 AP 有线接口。

在"网络端口信息"页面，单击"修改"按钮，按真实的 IP 地址、网关、DNS 来修改 AP 的 Ethernet 接口的状态参数，如图 10-2 所示。

图 10-2

步骤 7：配置虚拟 AP 和 SSID。

在"虚拟 AP 配置"页面根据需要配置 SSID，勾选"是否广播 SSID"，设置对应虚拟 AP 所关联的 VLAN 及采用的认证方式，最后选择"启用"复选框，如图 10-3 所示。

图 10-3

步骤 8：单击页面底部的"Update"按钮，保存配置，即完成操作。

经验分享

本实训中两个 SSID 对应两个 VLAN，所以需要在有线部分的三层设备（也是充当 AP 管理 VLAN 的网关和另外一个无线用户 STA_B 的网关）上配置路由策略保证无线用户与有线网络互通。如果配置两个 SSID 都关联到 VLAN1，则在配置第 2 个 SSID 时可以省略掉任务实施的第一部分。

实训 2　开放式无线接入配置

实训目标

信息部门使用 AC 完成了对 AP 瘦模式初始化操作和基础管理配置。市场部门因有特别要求，无线 AP 需要工作在胖模式，以方便市场部用户接入 Internet。

配置要求：无线用户开放式接入无线网络并能够接入互联网。无线网络的属性在 AP 上配置，无线用户的地址由有线交换机 DHCP 服务提供。

通过学习本实训，能够掌握无线胖模式概念，具备配置无线胖模式功能的能力。

实训设备

1）DCWL-8200AP 1 台。
2）CS6200 三层交换机 1 台，双绞线 2 条。
3）PC 2 台。

实训拓扑

实训拓扑图如图 10-4 所示。

图 10-4

实训任务

无线 AP 工作在胖模式，类似于家庭用无线宽带路由器无线配置，主要完成有线和无线两部分的配置、设置 SSID、配置加密方式、设置 DHCP 为客户端自动分配 IP 地址。配置步骤如下。

步骤 1：在 Ethernet 接口配置主菜单中执行 Manage → Ethernet Settings 命令。

根据网络的实际环境，配置 AP 的 IP 地址、所在 VLAN、网关地址、DNS 等参数，以保证 AP 能够与有线网络互通。配置参数如图 10-5 所示。配置完成后单击 Update 按钮保存。

经验分享

注意请启用本征 VLAN，并把 AP 与有线网络互连 VLAN 指定为本征 VLAN。

图 10-5

步骤 2：无线参数配置。

进入"射频参数配置"页面，"无线模式"和"信道"配置可以保持默认值，配置完成单击"Update"按钮保存配置。配置参数如图 10-6 所示。

图 10-6

步骤 3：配置并启用虚拟 AP。

单击左侧主菜单中的"虚拟 AP"，进入"虚拟 AP 配置"界面，如图 10-7 所示。此处配置只广播一个 SSID，名为"DCN_VAP_2G"，选择"是否广播 SSID"复选框表示对外广播 SSID；选择"启用"复选框，"VLAN 号"为默认的"1"，与 AP 管理地址属于同一个 VLAN。"安全设置"选择"None"。

图 10-7

配置完成后，单击"Update"按钮保存配置。查看 PC 的无线连接，发现此 WLAN 信号，如图 10-8 所示。

图 10-8

实训 3 无线加密接入的配置

实训目标

信息部门部署了无线局域网，受到公司员工欢迎。为了加强公司网络的安全性，需要提高无线网络接入身份验证机制。

配置要求：对上一实训中所配置的网络进行加密，要求选择三种加密方式之一，无线用户接入时先进行身份验证。

通过学习本实训，能够掌握无线加密方式，具备配置无线加密接入功能的能力。

实训设备

1）DCWL-8200AP 1 台。
2）CS6200 三层交换机 1 台，双绞线 2 条。
3）PC 2 台。

实训拓扑

实训拓扑图如图 10-9 所示。

图 10-9

实训任务

> **知识链接**
>
> 成熟的加密机制有 WEP 和 WPA 两大类。
>
> 1）WEP（Wired Equivalent Privacy，有线等效加密），通用于有线和无线网络加密。因为无线网络是用无线电把信息传播出去的，所以它特别容易被窃听。
>
> 2）WPA（Wi-Fi Protected Access，Wi-Fi 网络安全接入），有 WPA 和 WPA2 两个标准，是一种保护无线计算机网络（Wi-Fi）安全的系统，WPA2 具备完整的体系，WPA 或 WPA2 一定要启动并且被选来代替 WEP 才有用，但是在大部分的安装指引中，WEP 标准都是默认选项。

1. WEP 加密方式配置

步骤 1：在"虚拟 AP 配置"界面中设置"SSID"为"DCN-VAP-2G"，选择加密方式为"Static WEP"，如图 10-10 所示。

图 10-10

步骤 2：在 WEP 加密下方的窗口中，选择密钥格式。可以选择 64bits 加密为 5 位 ASCII 码或 10 位十六进制数，128bits 加密为 13 位 ASCII 码或 26 位十六进制数。

步骤 3：在 WEP Keys 中，输入使用的密码，单击"Update"按钮保存配置。

2．WPA 加密方式配置

步骤 1：在"虚拟 AP 配置"页面中可选择"安全设置"为"WPA Personal"，如图 10-11 所示。

图 10-11

步骤 2：如果在图 10-11 中选择"安全设置"为"WPA Enterprise"，则此方式需要 Radius 认证，在图 10-12 中输入 Radius 服务器地址，单击"Update"按钮保存配置。

图 10-12

3．802.1x 认证

802.1x 协议是基于 CLIENT/SERVER 的访问控制和认证协议。它可以限制未经授权的用户 / 设备通过接入端口访问 LAN/MAN。在获得交换机或 LAN 提供的各种业务之前，802.1x 对连接到交换机端口上的用户 / 设备进行认证。

选择 802.1x 认证，要求输入 Radius 服务器地址及 Radius 密钥，如图 10-13 所示。

图 10-13

第4部分

防火墙实训

第 11 章 防火墙技术基础

实训 1 防火墙透明模式的初始配置

实训目标

防火墙

防火墙是基础网络安全设备,是网络安全的第一道墙,从本实训起将学习防火墙的相关知识及实训。

在不改变目前网络结构的前提下,设置防火墙,把防火墙部署在网络中,对公司网络进行安全保护。

配置要求:

1)防火墙 ethernet0/2 接口和 ethernet0/3 接口配置为透明模式。

2)ethernet0/2 与 E0/3 同属于一个虚拟桥接组,ethernet0/2 属于 l2-trust 安全域,ethernet0/3 属于 l2-untrust 安全域。

3)为虚拟桥接组 vswitch1 配置 IP 地址以便管理防火墙。

4)允许网段 A ping 网段 B 及访问网段 B 的 Web 服务。

通过学习本实训,能够了解防火墙透明模式工作原理,具备配置透明模式功能的能力。

实训设备

1)DCFW-1800E-N3002 1 台。
2)CS6200 交换机 2 台,双绞线 2 条。
3)PC 6 台。

实训拓扑

实训拓扑图如图 11-1 所示。

图 11-1

实训任务

> **温馨提示**
>
> 本任务中防火墙的 vswitch 接口的 IP 地址与默认的 ethernet0/0 地址有冲突。如果使

用 WebUI 界面管理防火墙，则需要修改 ethernet0/0 接口 IP 地址为其他网段。

步骤 1：接口配置。

把 ethernet0/2 接口加入二层安全域。通过默认的 ethernet0/0 管理接口登录防火墙的 WebUI 管理页面，执行"配置"→"网络"→"网络连接"命令，页面的右侧显示出当前防火墙的所有接口状态，如图 11-2 所示。

图　11-2

图 11-2 页面下方是当前存在的物理接口的状态。选择需要配置的接口 ethernet0/2，单击上方的"编辑"按钮，在打开的"接口配置"页面中把 ethernet0/2 接口加入二层安全域 l2-trust，如图 11-3 所示。

图　11-3

把接口 ethernet0/3 加入二层安全域 l2-untrust，表示二层非信任区。注意，加入二层安全域的接口不能配置 IP 地址。

步骤 2：配置虚拟交换机（vswitch）。

选择 vswitchif1 接口，将其加入三层安全域 trust，配置管理 IP 地址为 192.168.1.254，如图 11-4 所示。

图　11-4

经验分享

此实训预留了防火墙的管理接口，才可以使用 WebUI 对防火墙进行管理。如果实际广泛应用中没有单独接口做管理，那么可以先使用控制线通过控制口登录防火墙，使用命令行配置：

DCFW-1800(config)# interface ethernet0/2
DCFW-1800(config-if-eth0/6)#zone l2-trust
DCFW-1800(config)# interface vswitchif1

```
DCFW-1800(config-if-vsw1)# zone trust
DCFW-1800(config-if-vsw1)# ip address 192.168.1.254/24
```

步骤 3：添加地址簿对象。

执行防火墙主页面右上方的"系统对象"→"地址簿"命令，在弹出的"地址簿"对话框上方单击"新建"按钮，定义网段 A（192.168.1.1 ~ 192.168.1.100）和网段 B（192.168.1.101 ~ 192.168.1.200）两个地址段对象。因为这里都不是一个完整的网段，所以选择定义时的 IP 范围，如图 11-5 和图 11-6 所示。

图 11-5

图 11-6

步骤 4：建立服务组对象，创建一个服务组，命名为"A-to-B_Service"，包括"PING"和"HTTP"两种服务，如图 11-7 所示。

图 11-7

安全策略配置如图 11-8 所示。

图 11-8

防火墙的透明模式初始配置已经完成，既起到了流量管理的作用，又没有影响到网络的架构，对用户来说，防火墙的存在是透明的。

实训 2　防火墙路由模式的初始配置

防火墙只有工作在路由模式才能够起到智能网关的作用，在内网与外网之间起到安全防范作用。当防火墙工作在路由模式时，需要对网络拓扑进行修改（内部网络用户需要更改网关、路由器需要更改路由配置等），此时相当于一台路由器。防火墙的 trust 区域接口与公司内部网络相连，untrust 区域接口与外部网络相连。采用路由模式时，可以完成 ACL 包过滤、ASPF 动态过滤、NAT 转换等功能。然而，在路由模式下这是一件相当费时的工作，因此在使用该模式时需权衡利弊。

实训目标

某公司作为一家高科技公司，网络信息安全非常重要。信息部门决定在网络边界的防火墙上启用智能网关功能，控制外网对内网数据的访问，防范外网的安全威胁，从而提高内网安全指数。

配置要求：

1）防火墙内网接口 ethernet0/0 加入 trust 安全域，使用默认的地址 192.168.1.1/24，外网接口加入 untrust 域，使用公网地址 222.1.1.1/24。

2）防火墙内网接口 ethernet0/0 开启 DHCP 和 DNS 代理服务，为内网用户提供 IP 地址和 DNS 代理。

通过学习本实训，能够了解防火墙路由模式工作原理，具备配置路由模式功能的能力。

实训设备

1）DCFW-1800E-N3002 1 台。

2）CS6200 交换机 2 台，双绞线 2 条。

3）PC 6 台。

实训拓扑

实训拓扑图如图 11-9 所示。

图 11-9

实训任务

步骤 1：接口地址配置。

在防火墙主页面中执行"网络"→"网络连接"命令，出现如图 11-10 所示的接口配置页面。

图 11-10

选择要配置的接口"ethernet0/1"，单击"编辑"按钮，弹出"接口配置"对话框，设置"绑定安全域"为"三层安全域"，然后将"安全域"设置为"untrust"，在"IP 配置"中选择"静态 IP"，然后输入为其分配的 IP 地址和网络掩码，如图 11-11 所示。

图 11-11

步骤 2：配置防火墙 DHCP 功能。

在图 11-11 所示的接口配置页面上，选中"etherent0/0"接口，单击"编辑"按钮进入编辑页面，单击"DHCP"按钮，然后单击"新建"按钮，打开如图 11-12 所示的页面，选择"DHCP 服务器"，设置网关、网络掩码、DNS 后添加地址池地址，单击"确定"按钮。

图　11-12

如果需要为某个主机指定 IP 地址，则可以选择图 11-12 中的"地址绑定"选项卡，输入要绑定的 IP 地址与 MAC 地址，所绑定的 IP 地址必须在地址池中，如图 11-13 所示。

图　11-13

步骤 3：验证。内网 PC 使用自动获取 IP 地址的方式来获取 IP 地址，可以看到 PC 已经获取到 192.168.1.111 的 IP 地址网关为 192.168.1.1，实现了 IP 地址绑定，如图 11-14 所示。

图　11-14

步骤 4：添加路由。

添加到外网的默认路由，允许访问所有的外部网络，下一跳地址为防火墙外网口的网关地址,这个地址应该在 ISP 设备上。访问防火墙主页面的菜单,执行"网络"→"路由"→"目的路由"命令,新建一条默认路由条目并添加下一跳,参数设置如图 11-15 所示。

图 11-15

步骤 5:添加源 NAT 策略。

内网用户众多,都使用私有地址,在访问外网时,使用源 NAT 转换,所有的私有地址都转换为防火墙外网接口 ethernet0/1 的公网地址 222.1.1.2。在防火墙主页面左侧执行"网络"→"NAT"→"源 NAT"命令添加源 NAT 策略,如图 11-16 所示。

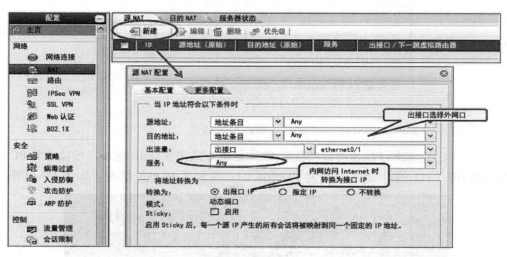

图 11-16

步骤 6:添加内网到外网的安全策略。

内网处于三层 trust 安全域,外网处于三层 untrust 安全域,防火墙默认不同安全域之间是不能访问的。在此为了让内网用户能够顺利访问外网,建立 trust 到 untrust 的安全策略,放行内网到外网的流量。

执行"安全"→"策略"命令,新建一条策略,选择从 trust 到 untrust,放行所有数据流量,如图 11-17 所示。

图 11-17

至此，防火墙的路由模式初始配置完成，内网用户可以访问外网，外网不能访问内网。既保证了内网用户的上网需求，又隐藏了内网的数据与信息。

实训 3　防火墙混合模式的初始配置

实训目标

某公司的 Web 服务器要对外提供服务，供 Internet 用户访问。要求配置防火墙既能作为内网用户的智能网关，又能保证 Web 服务器被外网用户安全访问。

某公司网络拓扑从防火墙的角度看，主要有三个区域，分别是内网（trust）、外网（untrust）和对外服务器（DMZ）。防火墙既需要为内网用户提供网关服务，又需要提供路由功能；同时防火墙的外网接口和服务器区的接口在同一网段，这两个接口必须工作在透明模式下。只有防火墙混合模式可以满足这个需求，内网口作为内网的网关，工作在路由模式，外网口和服务器接口共同作为桥接，工作在透明模式。

配置要求：

1）将 ethernet0/0 口设置成路由接口，ethernet0/1 和 ethernet0/2 口设置成二层接口，并设置 vswitch 接口。

2）设置源 NAT 策略。

3）配置安全策略。

通过学习本实训，能够了解防火墙混合模式工作原理，具备配置混合模式功能的能力。

实训设备

1）DCFW-1800E-N3002 1 台。

2）CS6200 交换机 2 台，双绞线 2 条。

3）PC 6 台。

实训拓扑

实训拓扑图如图 11-18 所示。

图 11-18

实训任务

步骤 1：配置接口地址。

内网口 ethernet0/0 口设置成三层安全域 trust，配置默认的地址 192.168.1.1/24，如图 11-19 所示。

图 11-19

设置外网口，ethernet0/1 接口连接 Internet，设置成二层安全域 l2-untrust，如图 11-20 所示。

图 11-20

设置外网口，ethernet0/2 接口连接服务器，设置成二层安全域 l2-dmz，如图 11-21 所示。

第 11 章 防火墙技术基础

图 11-21

步骤 2：配置 vswitch 接口。

由于二层安全域接口不能配置 IP 地址，需要将地址配置在网桥接口上，该网桥接口为 vswitchif 1。vswitchif 1 接口的地址配置如图 11-22 所示。

图 11-22

步骤 3：设置源 NAT 策略。

针对内网所有地址的上网需求，需要在防火墙上设置源 NAT，内网 PC 在访问外网时，凡是从 vswitch 接口出去的数据包都做地址转换，转换地址为 vswitch 接口地址，如图 11-23 所示。

图 11-23

步骤 4：添加路由。

要创建一条到外网的默认路由，如果内网有三层交换机的话，则还需要创建到内网的回指路由，这里默认路由的下一跳地址设为网关 218.240.143.1，如图 11-24 所示。

图 11-24

步骤 5：设置地址簿。

在配置放行安全策略时，需要选择相应的地址和服务进行放行，所以这里首先要创建服务器的地址簿。在创建地址簿时，如果创建的服务器属于单个 IP，则建议使用 IP 成员方式，掩码一定要写 32 位，如图 11-25 所示。

图 11-25

步骤 6：配置放行策略。

配置放行策略时，首先要保证内网能够访问到外网。放行到内网口所属安全域到 vswitch 接口所属安全域的安全策略，应该是从 trust 到 untrust，参数设置如图 11-26 所示。

图 11-26

另外还要保证外网能够访问 Web_Server，该服务器的网关地址设置为 ISP 网关 218.240.143.1，需要放行二层安全域之前的安全策略，应该是放行 l2-untrust 到 l2-dmz 的策略。

实训 4　防火墙 SNAT 配置

由于公司内部私网地址较多，运营商只分配给一个或者几个公网地址。在这种条件下，这几个公网地址需要满足几十乃至几百几千人同时上网时，需要配置 SNAT。

1）防火墙的接口名称。在防火墙中，WAN 接口是防火墙用来连接外网的接口。LAN 接口是外网接口，DMZ 接口用来连接服务器，其他防火墙的接口对应关系在面板上可以找到。

2）SNAT，源地址转换，它的作用是将 IP 数据包的源地址转换成另外一个地址。内部地址要访问公网上的服务时（如 Web 访问），内部地址会主动发起连接，由防火墙上的网关对内部地址做地址转换，将内部地址的私有 IP 转换为公网的公有 IP，防火墙网关的这个地址转换称为 SNAT，主要用于内部共享 IP 访问外部。

通过学习本实训，能够掌握防火墙 SNAT 概念，具备配置 SNAT 功能的能力。

实训目标

考虑到公网地址有限，不能每台 PC 都能配置公网地址访问外网。通过少量公网 IP 地址来满足多数私网 IP 上网，以缓解对 IP 地址需求的压力。

实训设备

1）DCFW-1800E-N3002 1 台。
2）CS6200 交换机 2 台，双绞线 2 条。
3）PC 6 台。

实训拓扑

实训拓扑图如图 11-27 所示。

图　11-27

配置防火墙使内网 192.168.1.0/24 网段可以访问 Internet。

实训任务

步骤 1：配置接口。
首先通过防火墙默认 ethernet0/0 接口地址 192.168.1.1 登录防火墙界面进行接口的配置。

通过 WebUI 登录防火墙界面，如图 11-28 所示。

图 11-28

输入默认用户名 admin、密码 admin 后单击"登录"按钮，配置外网接口地址。参数设置如图 11-29 所示。

图 11-29

步骤 2：添加路由。

添加到外网的默认路由。执行"网络"→"路由"→"目的路由"命令新建路由条目，添加下一跳地址，如图 11-30 所示。

图 11-30

第 11 章　防火墙技术基础

步骤 3：添加源 NAT 策略。

执行"防火墙"→"NAT"→"源 NAT"命令添加源 NAT 策略，如图 11-31 所示。

图　11-31

步骤 4：添加安全策略。

执行"防火墙"→"策略"命令，选择好"源安全域"和"目的安全域"后，新建策略，如图 11-32 所示。

图　11-32

关于源 NAT，只需要建立一条内网口安全域到外网口安全域放行的策略就可以保证内网能够访问到外网。如果需要对策略中各个选项有更多的配置要求则可以进行策略高级配置，如图 11-33 和图 11-34 所示。

图　11-33

图　11-34

实训 5 防火墙 DNAT 配置

由于公网地址有限，一般在申请线路时，运营商分配的只有一个或几个公网地址。但是内部服务器设置成私网地址后，需要将私网地址映射到公网，外网用户才可以通过映射后的公网地址访问到服务器。映射包括两种：一种为端口映射，端口映射需要的服务器端口；另一种为 IP 映射，将私网地址和公网地址做一对一的映射。

1）DMZ "隔离区"。此 "隔离区" 也称 "非军事化区"。它是为了解决安装防火墙后外部网络不能访问内部网络服务器的问题，而设立的一个非安全系统与安全系统之间的缓冲区，这个缓冲区位于企业内部网络和外部网络之间的小网络区域内，在这个小网络区域内可以放置一些必须公开的服务器设施，如企业 Web 服务器、FTP 服务器和论坛等。另一方面，通过这样一个 DMZ 区域，更加有效地保护了内部网络，因为这种网络部署比起一般的防火墙方案，对攻击者来说又多了一道关卡。

2）DMZ 访问控制策略。①内网可以访问外网；②内网可以访问 DMZ；③外网不能访问内网；④外网可以访问 DMZ；⑤ DMZ 不能访问内网；⑥ DMZ 不能访问外网。

3）DNAT。目的地址转换的作用是将一组本地内部的地址映射到一组全球地址，也就是对数据包的源地址和目的地址进行修改，并且保存修改前后的映射关系，根据需要进行还原操作。

4）多对一的地址转换的限制。为了解决 IP 地址不足的问题，神州数码防火墙允许许多内部的机器对应到一个合法的 NIC IP 地址，使得这些内部机器可同时用一个合法 NIC IP 连上互联网。但是从互联网无法传递一个连接到内部的机器，因为防火墙不知道互联网传来的连接传给哪一部机器。如果内部有服务器对外提供服务，则需要一对一的地址转换。

实训目标

防火墙上配置了 SNAT 后，内部用户在访问外网时都隐藏了私网地址。如果防火墙内部有一台服务器需要对外网用户开放，则此时就必须在防火墙上配置 DNAT，将数据包在防火墙做目的地址转换，让外网用户访问到该服务器。

通过学习本实训，能够掌握防火墙 DNAT 概念，具备配置 DNAT 功能的能力。

实训设备

1）DCFW-1800E-N3002 1 台。

2）CS6200 交换机 2 台，双绞线 2 条。
3）PC 6 台。

实训拓扑

实训拓扑图如图 11-35 所示。

图 11-35

实训任务

步骤 1：外网口 IP 为内网 FTP Server 及 Web ServerB 做端口映射，并允许外网用户访问该 Server 的 FTP 和 Web 服务，其中 Web 服务对外映射的端口为 TCP 8000。

1）配置准备工作。设置地址簿，执行"对象"→"地址簿"命令设置服务器地址，如图 11-36 所示。

图 11-36

设置服务簿，防火墙出厂自带一些预定义服务，但是如果需要的服务在预定义中不包含时，需要执行"对象"→"服务簿"命令自定义服务，如图 11-37 所示。

2）创建目的 NAT。配置目的 NAT，为 trust 区域 Server 映射 FTP（TCP 21）和 HTTP（TCP 80）端口，如图 11-38 所示。

图 11-37

图 11-38

3）创建放行安全策略。创建放行安全策略，允许 untrust 区域用户访问 trust 区域 Server 的 FTP 和 Web 应用，如图 11-39 所示。

图 11-39

步骤 2：允许内网用户通过域名访问 Web ServerB（即通过合法 IP 访问）。

实现这一任务所需要做的就是在之前的配置基础上，增加 trust → trust 的安全策略，如图 11-40 所示。

图 11-40

步骤 3：使用合法 IP_218.240.143.220 为 Web ServerA 做 IP 映射，允许内外网用户对该 Server 的 Web 访问。

1）配置准备工作。将服务器的实际地址使用"web_serverA"来表示，即添加地址对象，如图 11-41 所示。

图 11-41

将服务器的公网地址使用"IP_218.240.143.220"来表示,如图11-42所示。

图 11-42

2)配置目的NAT。创建目的NAT条目,单击"新建",选择IP映射,如图11-43所示。

图 11-43

3)放行安全策略。放行untrust区域到dmz区域的安全策略,使外网可以访问dmz区域服务器,如图11-44所示。

图 11-44

放行trust区域到dmz区域的安全策略,使内网机器可以用公网地址访问dmz区域内的服务器,如图11-45所示。

图 11-45

参 考 文 献

[1] STEVENS W R. TCP/IP 详解卷 1：协议 [M]. 吴英，张玉，许昱玮，译. 北京：机械工业出版社，2016.
[2] DOYLE J. TCP/IP 路由技术：第 2 卷 [M]. 夏俊杰，译. 北京：人民邮电出版社，2017.
[3] 沈鑫剡，魏涛，邵发明，等. 路由和交换技术 [M]. 2 版. 北京：清华大学出版社，2018.